Handbook of Thin Film Process Technology

Editors

David A Glocker
Isoflux Inc., NY, USA

S Ismat Shah
E I du Pont de Nemours & Co., Wilmington, DE, USA

98/1 Reactive Sputtering

Guest Editor

William D Westwood
Formerly of Bell Northern Research, Ontario, Canada

CRC Press
Taylor & Francis Group
Boca Raton London New York

CRC Press is an imprint of the
Taylor & Francis Group, an **informa** business

First published 1998 by CRC Press
Taylor & Francis Group
6000 Broken Sound Parkway NW, Suite 300
Boca Raton, FL 33487-2742

Reissued 2018 by CRC Press

British Library Cataloguing-in-Publication Data

A catalogue record for this book is available from the British Library.

ISBN 0 7503 0528 2

Library of Congress Cataloging-in-Publication Data are available

A Library of Congress record exists under LC control number: 2008530032

Publisher's Note
The publisher has gone to great lengths to ensure the quality of this reprint but points out that some imperfections in the original copies may be apparent.

Disclaimer
The publisher has made every effort to trace copyright holders and welcomes correspondence from those they have been unable to contact.

ISBN 13: 978-1-315-89368-6 (hbk)
ISBN 13: 978-1-351-07278-6 (ebk)

Visit the Taylor & Francis Web site at http://www.taylorandfrancis.com and the
CRC Press Web site at http://www.crcpress.com

How to use the Supplement

This package contains the fifth supplement (98/1: Reactive Sputtering) to the Handbook of Thin Film Process Technology. It can be used as a stand-alone product but, if you find it useful in your work, you may be interested in the main Handbook (see special offer below).

For those purchasers who also own the main Handbook, the supplement is hole-punched to allow it to be incorporated into the Handbook binder. This supplement on Reactive Sputtering is completely new and should be inserted into the Handbook following section A4. The main contents list, the list of contributors and the index have all been updated to take account of the new material, and are also included in this supplement. Handbook owners can remove the original items from the Handbook and replace them with the new versions.

Contents

A Physical Deposition Techniques

A1 Thermal Evaporation
Coordinating Editors: *E B Graper and J Vossen*

A1.0 Introduction and general discussion
E B Graper

A1.1 Resistance evaporation
E B Graper

A1.2 Electron beam evaporation
E B Graper

A1.3 Ion vapour evaporation
E B Graper

A1.4 Cathodic arc deposition
P J Martin

A1.5 Laser ablation
A Morimoto and T Shimizu

A2 Molecular Beam Epitaxy
Coordinating Editors: *S A Barnett and J Poate*

A2.0 Introduction and general discussion
S A Barnett and I T Ferguson

A2.1 Semiconductor growth by metalorganic molecular beam epitaxy (MOMBE)
C R Abernathy

A2.2 Gas-source MBE
G Y Robinson

A2.3 Chemical beam epitaxy
T H Chiu

A2.4 Thin film deposition and dopant incorporation by energetic particle sources
S Strite and H Morkoç

A3 Sputtering
Coordinating Editors: *S I Shah and D Glocker*

A3.0 Introduction and general discussion
S I Shah

A3.1 Glow discharge sputtering
A S Penfold

A3.2 Magneton sputtering
A S Penfold

A3.3 Ion-beam sputtering
T Itoh

A3.4 Primary ion beam deposition
K Miyake

A4 Thermal Spray Coatings
Coordinating Editor: *R C Tucker Jr*

A4.0 Introduction to thermal spray coatings
R C Tucker Jr

A4.1 Flame spray
P A Kammer

A4.2 Plasma spray coatings
R C Tucker Jr

A4.3 High velocity oxy-fuel coatings
R C Tucker Jr

A4.4 Detonation gun deposition
R C Tucker Jr

A4.5 Mechanical, wear, corrosion and other properties of thermal spray coatings
R C Tucker Jr

A5 Reactive Sputtering
Coordinating Editor: *W D Westwood*

A5.0 Reactive Sputtering: Introduction and General Discussion
W D Westwood

A5.1 Reactive pulsed DC magnetron sputtering and control
J M Schneider and W D Sproul

A5.2 AC and RF reactive sputtering
G O Este and W D Westwood

A5.3 Modeling of the reactive sputtering process
S Berg, T Nyberg, H-O Blom and C Nender

B Chemical Deposition Techniques

B1 Chemical Vapour Deposition
Coordinating Editor: *L Vescan*

B1.0 Introduction and general discussion
L Vescan

B1.1 Metalorganic chemical vapour deposition (MOCVD)
R D Dupuis

B1.2 Plasma-enhanced chemical vapour deposition
F Jansen

B1.3 Photoassisted chemical vapour deposition
S J C Irvine

B1.4 Thermally activated chemical vapour deposition
L Vescan

B1.5 Atomic layer epitaxy
T Suntola

C Processing Technologies

C1 Pattern Transfer
Coordinating Editors: *J W Coburn*

C1.0 Introduction and general discussion
J W Coburn

C1.1 Reactive ion etching
C Steinbrüchel

C1.2 Ion-beam-based chemical dry etching
C Steinbrüchel

C1.3 Ion milling
C Steinbrüchel

D Real-Time Diagnostics

D0 Introduction and General Discussion
Coordinating Editor: *R W Collins*

D1 Diagnostic techniques

D1.0 Reflection high-energy electron diffraction as a diagnostic technique
B A Joyce

D1.1 Low-energy electron diffraction
Sheng-Liang Chang and P A Thiel

D1.2 Reflection mass spectrometry
R Kaspi

D2 Optical Diagnostics

D2.0 Infra-red emission interferometry
A J SpringThorpe

D2.1 Reflectance anisotropy
B Drévillon

D2.2 Interferometry as an *in situ* probe during processing of semiconductor wafers
V M Donnelly

D2.3 Ellipsometry
P Snyder

D2.4 Photoluminescence
P R Berger

D2.5 Elastic laser light scattering
B Gallois

D2.6 Infrared diode laser absorption spectroscopy as a plasma chemical diagnostic
R C Woods

D3 Plasma Probes

D3.0 Langmuir probe diagnostics
N Hershkowitz

D3.1 Microwave interferometers
R A Breun

D3.3 Atomic absorption spectroscopy
Chih-shun Lu

D4 Other Diagnostics
Coordinating Editor: *R Collins*

D4.0 Quartz monitors and microbalances
J Krim and C Daly

D4.1 Probes of film stress
D Glocker

E Surface Modification in Vacuum

Coordinating Editor: *D Glocker*

E1 Substrate Cleaning

E1.0 Processes for substrate cleaning
D Mattox

E2 Surface Treatment for Corrosion and Wear Protection

E2.1 Material aspects of corrosion protection
C M Cotell

E2.2 Ion implantation with beams
M Nastasi

E2.3 Plasma source ion implantation
D J Rej

E3 Surface Treatment of Polymers for Adhesion

E3.0 Plasma sources for polymer surface treatment
M R Wertheimer, L Martinu and E M Liston

E3.1 Surface chemistry of treated polymers
L Gerenser

F Superlattices and Multilayered Structures

F1 Growth Phenomena in Superlattices
D Jesson

F2 Mechanical Behaviour of Artificially Multilayered Thin Films
R C Cammarata

F4 Magnetic Superlattices
E Fullerton

F5 Multilayered Structures for X-ray Mirrors
E Spiller

F6 Superlattices for Optoelectronic Applications

F6.1 Si-based superlattices

F6.1.1 Photonic applications *F Y Huang and B Jalai*
F6.1.2 Electronic applications *J Cressler*

F6.2 Compound and alloy semiconductor superlattices
L Salamanca-Riba

F7 Self-Propagating Reactions in Multilayer Materials
T Weihs

X Materials

X1 Hard and Protective Materials

X1.0 Introduction: *O Knotek and A Schrey*
X1.1 TiN
X1.2 TiAlN
X1.3 TiAlVN
X1.4 CrN
X1.5 ZrN
X1.6 HfN
X1.7 BN
X1.8 Diamond
X1.9 Ni–Cr–B–Si
X1.10 Al-bronze
X1.11 Al_2O_3–TiO_2

X2 Electronic Materials

X2.0 Introduction: *K Cadien and S Sivaram*
X2.1 GaAs
X2.2 a-Si:H
X2.3 AlGaAs
X2.4 Tellurides
X2.5 $CuInSe_2$
X2.6 Si
X2.7 Ge
X2.8 Si–Ge
X2.9 W
X2.10 GaN
X2.11 AlN
X2.12 ErAs
X2.13 Quaternaries
X2.14 Silicides
X2.15 SiSnC
X2.16 SiN

X3 Optical Materials

X3.0 Introduction: *J Targove*
X3.1 AlN
X3.2 ZnO
X3.3 $PbTiO_3$
X3.4 $KNbO_3$
X3.5 TiO_2
X3.6 In_2O_3:Sn

X4 Ferroelectric Materials

X4.0 Introduction: *M Sayer*
X4.1 $Bi_4Ti_3O_{12}$

X4.2 $LiNbO_3$ and $LiTaO_3$
X4.3 $PbTiO_3$/$PbZrTiO_3$

X5 Ferromagnetic Materials
X5.0 Introduction: *E M T Velu and D N Lambeth*
J Q Xiao
G Hadjipanayis
X5.1 CoCr
X5.2 TbFeCo
X5.3 CoPt/CoPd
X5.4 GdTbFe

X6 Superconducting Materials
X6.0 Introduction: *J Azoulay*
X6.1 NbN
X6.2 $YBa_2Cu_3O_7$
X6.3 Thallium-based compounds
X6.4 Mercury-based compounds

X7 Miscellaneous Materials
X7.1 PTFE
X7.2 PPN
X7.3 Ir/Pt

Appendix A: List of Contributors

Subject Index

Part A
PHYSICAL DEPOSITION TECHNIQUES

A1 THERMAL EVAPORATION

Coordinating Editors: E B Graper, LeBow Company, Goleta, CA, USA
J Vossen (deceased)

A1.0 Introduction and general discussion
E B Graper

A1.1 Resistance evaporation
E B Graper

A1.2 Electron beam evaporation
E B Graper

A1.3 Ion vapour evaporation
E B Graper

A1.4 Cathodic arc deposition
P J Martin

A1.5 Laser ablation
A Morimoto and T Shimizu

A2 MOLECULAR BEAM EPITAXY

Coordinating Editors: S Barnett, Northwestern University, Evanston, IL, USA
J Poate, AT&T Bell Laboratories, Murray Hill, NJ, USA

A2.0 Introduction and general discussion
S Barnett and I T Ferguson

A2.1 Semiconductor growth by metalorganic molecular beam epitaxy (MOMBE)
C R Abernathy

A2.2 Gas-source MBE
G Y Robinson

A2.3 Chemical beam epitaxy
T H Chiu

A2.4 Thin film deposition and dopant incorporation by energetic particle sources
S Strite and H Morkoç

A3 SPUTTERING

Coordinating Editors: S I Shah, E.I. DuPont de Nemours, Wilmington, DE, USA
D Glocker, Eastman Kodak Research Laboratories, Rochester, NY, USA

A3.0 Introduction and general discussion
S I Shah

A3.1 Glow discharge sputtering
A S Penfold

A3.2 Magneton sputtering
A S Penfold

A3.3 Ion-beam sputtering
T Itoh

A3.4 Primary ion beam deposition
K Miyake

A4 THERMAL SPRAY COATINGS

Coordinating Editor: R C Tucker Jr, Praxair Surface Technologies Inc., Indianapolis, IN 46224, USA

A4.0 Introduction to thermal spray coatings
R C Tucker Jr

A4.1 Flame Spray
P A Kammer

A4.2 Plasma spray coatings
R C Tucker Jr

A4.3 High velocity oxy-fuel coatings
R C Tucker Jr

A4.4 Detonation gun deposition
R C Tucker Jr

A4.5 Mechanical, wear, corrosion and other properties of thermal spray coatings
R C Tucker Jr

A5 REACTIVE SPUTTERING

Coordinating Editor: W D Westwood, formerly of Bell Northern Research, Ontario, Canada

A5.0 Reactive sputtering: introduction and general discussion
W D Westwood

A5.1 Reactive pulsed DC magnetron sputtering and control
J M Schneider and W D Sproul

A5.2 AC and RF reactive sputtering
G O Este and W D Westwood

A5.3 Modeling of the reactive sputtering process
S Berg, T Nyberg, H-O Blom and C Nender

A5

Reactive Sputtering

A5.0 Reactive Sputtering: Introduction and General Discussion

William D Westwood

A5.0.0 INTRODUCTION

Reactive sputtering is a method to deposit films which have a different composition from the target by adding a gas to the sputtering system to produce a reaction of the gas with the target material. The reaction may be controlled to dope the film (e.g. nitrogen doped Ta films) or enough gas may be supplied to ensure complete reaction of the sputtered metal. In the former case, the main sputtering gas is Ar or another inert gas but the fully reacted film may be sputtered either in a similar gas mixture (e.g. Ar/N_2) or in the reactive gas alone. Reactive sputtering was discussed briefly at the end of section A3.0.

Reactive sputtering was probably first observed by Grove [1] in the initial discovery of the sputtering phenomenon; he sputtered Ag in residual air due to the poor vacuum available at the time. Reactive sputtering has often been carried out inadvertantly because of inadequate understanding of the reactivity in the discharges used for sputtering. However, the first real application of reactive sputtering was to deposit Ta_2N films for resistors in thin film hybrid circuits [2]. By controlling the flow rate of N_2, the electrical properties of the film was adjusted to the required value by doping the Ta. If the N_2 flow is too high, the completely reacted TaN is deposited. The sputtering of these resistor films was thoroughly covered in a book [3] but this predated the use of magnetron sputtering systems. The development of reactive sputtering methods and the range of applications have been reviewed [4, 5].

Reactive sputtering for film deposition was carried out initially by diode sputtering with simple DC power supplies. Both metallic (e.g. Ta) and non-metallic (e.g. $NiFe_2O_4$) targets were sputtered; the oxide ceramic targets were made sufficiently conducting by firing in a reducing atmosphere [6]. High quality dielectric films were deposited; e.g. Ta_2O_5 optical waveguides [7]. Although dielectric layers formed on the targets, they were sufficiently conducting for the low current densities of diode sputtering. However, when high current density magnetrons were used, serious arcing caused defects in the films. The arcing is caused by the charging of the insulating layers by the ion current in the plasma and the subsequent breakdown of the dielectric when the breakdown field of the dielectric is exceeded. When the current density is low, this charge is dissipated by the leakage current through the layer but the commercial requirement is for high deposition rates.

The solution to the problem of insulating targets was solved in the 1960s by the introduction of RF (typically 13.56 MHz) diode sputtering [8]. RF sputtering is discussed in A3.0.1. Due to the much higher mobility of electrons than ions, the target potential is established at $-V_0$ relative to the plasma, where V_0 is the amplitude of the RF voltage; the target can be positive relative to the plasma for only a small part of each RF cycle so that the positive (ions) and negative (electrons) charges reaching it over the complete RF cycle are equal, as required by the RF circuit. In fact, RF sputtering is ideally suited to two identical targets. In this 'symmetric' case, the processes at one electrode are identical to those at the other except for a phase difference; both targets are sputtered at the same rate. However, this 'symmetric' case was not well suited to the diode system geometries in use at the time. To make the RF diode system similar to the DC diode, the symmetry was broken by increasing the capacitance of one electrode relative to the other; this reduces the potential difference between the 'substrate electrode' and the plasma so that it approximates the anode in the DC diode.

The symmetric RF diode system was used, with the targets facing (or opposed), in cases where negative ions (e.g. O^-) are sputtered from the target (e.g. ZnO, high T_c oxides); the substrates are located off to the side to avoid the flux of highly energetic species [9].

RF power supplies were applied to magnetron targets and solved the problem of insulating targets becoming charged. However, it did not solve the other problem with many insulating targets; i.e. the intrinsically low sputter yield of some materials. For example, the yield of Al from Al_2O_3 is about 5% of the yield from Al.

Another problem with RF magnetron sputtering was explored by Este and Westwood [10]; the sputtering rate was only about 50% of that for DC or low frequency (< 30 kHz) sputtering because only 50% of the power registered by the RF supply was dissipated in the targets. Although a full explanation for this effect is not yet available, the loss of sputtering power at high frequency is real.

Thus, there are two basic problems with sputtering insulating targets or insulating regions or layers on metal targets; first, charging of the target layers; second, the low sputter yield for these materials. The latter may still occur if the compound layer on the target is a conductor; e.g. TiN is a good conductor but the yield of Ti is still only 25% of that from Ti.

Different approaches have been taken to try to solve these problems. In the following parts of this section, several techniques are described in detail; section A5.1 addresses the control issues involved with maintaining the target free of the compound using basically a DC power supply, although it is actually modulated in a step-wise function to minimize the effect of any arcing which may occur. This is essentially an extreme case of 'doping' the film; i.e. the reactive gas must be controlled at the correct level. In doping,such as for the TaN_x resistor films, the N_2 is controlled to give the correct value of x. In depositing a fully reacted compound like Ta_2O_5, the O_2 has to be controlled to prevent oxide formation on the target. The specification of 'fully reacted' is important. The Ta_2O_5 film can be visually transparent but still be oxygen deficient and thus have poor dielectric properties. However, a slight oxygen deficiency is advantageous in oxide films (e.g. SnO_2) used as transparent conductors.

Section A5.2 deals with the use of full wave modulation of the power to prevent target charging even when the conducting target is covered by an insulating layer. It is usually preferable to use a metal target; fabrication is simpler, and less expensive, than for a ceramic target which involves calcining, pressing and firing. The metal target is also likely to have a lower impurity content and is easier to cool because of the higher thermal conductance. However, there are some materials (e.g. $LiNbO_3$) for which a metallic target is not feasible; the sputter yields are typically low in these cases. Full wave modulation may be either AC ($\sim$ 40 kHz) or RF (13.56 MHz) frequencies.

The most appropriate method will depend on the application. A major consideration is often the deposition rate which can be obtained. Table A5.0.1 gives the maximum possible deposition rates for some examples, assuming that the erosion and deposition areas are identical. These calculations are based upon the sputter yields for the elements and the measured ratios of deposition rates of metal and compound from a target covered by the compound layer. The rate per W for the element (row 5) is calculated from the elemental yields for 500 eV Ar^+ sputtering (row 1); the current density is 2 mA cm^{-2}. As an example, for a typical circular planar magnetron with an erosion track area of 50 cm^2 (i.e. uniform sputtering from a track with inner and outer radii of 3 and 5 cm), the erosion rate per kW of sputtering power is given in row 6. If the reactive gas simply added to the element and the deposition and erosion areas were equal, the compound deposition rates in row 7 of the table would be obtained; i.e. the element is sputtered from the target by Ar^+ and the atoms react at the substrate. The deposition rate therefore increases.

These deposition rates cannot be achieved in a real system; first, the elemental sputtering rate decreases since some reactive gas atoms reach the target and are sputtered instead of the element; second, the sputtered atoms deposit over a larger area than the target erosion area. The latter can be calculated [11, 12]; for a distance of 5 cm, the rate is only 7% of these erosion rates and decreases to 3.4% for 10 cm and to 1.8% for 15 cm. The maximum possible deposition rates on a substrate at a distance of 10cm are given in row 8. If substrates are moved during deposition to improve deposition uniformity, the average rates are even lower. Of course, the actual deposition rate can be higher if the sputtering power is increased.

Experimentally, the measured deposition rates usually increase initially as the reactive gas is added and then decrease rapidly as the compound layer forms on the target. The values in table A5.0.1 for the compound rate in this condition (row 9) were determined relative to the elemental rates from deposition

Table A5.0.1. Maximum rates for metal and compound layers. Rows 1–5 are calculated from the elemental yields and row 6 is the erosion rate for a circular magnetron and row 7 the maximum deposition rate on a substrate parallel to the target; row 8 are values for the compound by adding the reactive gas atoms to the element; row 9 is the experimental deposition rate for a 'poisoned' target; rows 10–12 are the individual yields from the compounds.

row		Al	Si	Ti	Nb
1	Yield for 500 eV Ar^+ (at/ion)	1	0.5	0.5	0.7
2	10^{15} atoms cm^{-2} s^{-1} per mA	6.25	3.12	3.12	4.38
3	μg cm^{-2} s^{-1} per mA	0.28	0.15	0.25	0.68
4	nm s^{-1} per mA	1.05	0.62	0.55	0.78
5	nm s^{-1} per W	2.1	1.24	1.1	1.56
6	rate (nm s^{-1}) for 1 kW over 50 cm^2 erosion track	42	24.8	22	31.2
7	max. dep. rate at 10 cm spacing (nm s^{-1} kW^{-1})	1.4	0.85	0.75	1.06
	compound	Al_2O_3	SiO_2	TiN	Nb_2O_5
8	maximum dep. rate at 10 cm spacing (nm s^{-1} kW^{-1})	2.5	1.9	0.8	2.8
9	measured rate relative to element (nm s^{-1} kW^{-1})	0.07	0.34	0.51	0.36
	Yields from compound (at/ion)				
10	Metal	0.034	0.08	0.11	0.027
11	O	0.051	0.16		0.067
12	N			0.11	

rate measurements; they may differ slightly from system to system. These rates assume the same erosion area as for the elemental targets.The rates which can be obtained from a compound target, or the compound layer on the target, are much lower than can be obtained from the metal target; e.g. for Al_2O_3, the measured rate is 5% of the deposition rate for Al (0.07 cf 1.4 nm s^{-1}), at the same sputtering power. This is the motivation for the many approaches to deposit compound films from an elemental target without formation of the layer on the target. These experimentally determined values can be used to calculate the approximate sputtering yields from the target (rows 10–12).

Thus, for the deposition of Al_2O_3 in the example of table A5.0.1, the maximum possible rate on a substrate is 2.5 nm s^{-1} kW^{-1} from an unpoisoned target and 0.07 nm s^{-1} kW^{-1} for a poisoned target. Reported rates for unpoisoned targets [13, 14] are typically only $\sim$ 0.2 nm s^{-1} kW^{-1} whereas the values for poisoned targets are similar to those in the table.

The minimum reactive gas flows required can also be calculated from the target erosion rate. Assuming that all the O_2 reacts with the sputtered Al, in table A5.0.1 for example, requires a gas flow of 21.5 sccm. Since some of the O_2 will be pumped away without reaction, higher flow rates are required; the higher the system pumping speed, the higher the required flow.

Since many of the systems for reactive sputtering pump O_2 through a cryopump, a word of caution is in order [15]. The unused O_2 is condensed on the pump along with species generated in the sputtering plasma; ozone may be one of these. Thus, care must be taken in regenerating the cryopump since rapid desorption of O_2 and the very reactive O_3 could have serious consequencies. Frequent regeneration is necessary to minimize the build-up and purging the system with N_2 during regeneration with an appropriate exhaust for O_3 should be used.

Section A5.3 outlines some approaches to modelling the reactive sputtering processes to better define the possible operational parameters as a guide to experiments for the deposition of both simple (e.g. AlN) and complex (e.g. $PbTi_{1-x}Zr_xO_3$, SiO_xN_y) compounds.

In the rest of this section, some of the issues in reactive sputtering are discussed and other deposition methods are summarized.

A5.0.1 APPLICATIONS

The initial use of reactive sputtering for depositing TaN_x thin film resistors has continued but the range of applications has greatly increased. Almost any conceivable combination of targets, reactive gasses, and sputtering systems has been used to deposit films on all types and sizes of substrates.

Transparent conducting oxides, which are oxygen deficient semiconductors, are sputtered for solar cell contacts, display panels and de-icing windshields for cars and airplanes. Similar materials, such as ZnO, are also deposited for temperature control coatings. These metals are not too reactive and can be sputtered at reasonably high rates [16]. Oxide films are used as protective layers on plastic for automobile applications [17]. Nitride layers and multilayers are commonly used for hard coatings on tools, etc [18]. Mattox [19] reviewed the properties of oxide, nitride and carbide materials for hard coatings.

The most widely used films for diffusion barriers in electronics applications are TiN or (Ti:W)N although other materials are increasingly being used for special applications. TaSiN films provide a barrier to oxygen diffusion [20]. Ta(N,O) films [21] and amorphous $WSi_{0.6}N$ [22] are barriers to Cu diffusion.

Films with good magnetic and ferroelectric properties have been studied [23] for applications as high storage density capacitors (e.g. in DRAMs) and magnetic storage. However, it is very difficult to obtain the desired electrical and magnetic properties, even if the films appear structurally and crystallographically correct.

Films are used as optical coatings in many areas,including solar control for buildings and cars, flat panel displays, and architectural and decorative coatings. Ion assisted processes for optical coatings were reviewed by Martin and Netterfield [24] and Dobrowolski *et al* [25] reviewed oxide films for interference filters.

ZnO is a piezoelectric film which is used in SAW filters. AlN has also been used for SAW transducers and for high temperature strain gauges [26]. $LiNbO_3$ and $LiTaO_3$films are desirable for optical modulation but the epitaxial films required are difficult to obtain by sputtering [27, 28]

The requirements of the films depend very much on the application; e.g. an Al_2O_3 film for a protective coating has quite different requirements from a film for optical coatings or dielectric applications. This distinction is not always made clear in the literature; 'good' films for one application may be inadequate for another, even if the applications are superficially similar; e.g. a single thin dielectric layer may be sufficient for an architectural glass coating, a laser mirror or a capacitor but the 'quality' requirements are significantly different. For any application, the best compromise must be made in selecting the deposition conditions.

A5.0.2 DOPING

Reactive sputtering was developed in the 1950s to deposit the doped Ta-N films for hybrid circuits. By controlling the flow rate of N_2, the electrical properties of the film was adjusted to the required value. Many papers report these results in terms of the partial pressure of the reactive gas. If the partial pressure is measured prior to sputtering, the value will be lower during deposition because the gas reacts with the sputtered metal; the value will also vary with the pumping speed for a given flow rate . For partial pressure to be a useful metric, it must be measured during deposition, by a residual gas anlyzer (RGA), for example.

Doping depends on the reactive gas flow in a complex way. For example, although the N content in the film is proportional to the N_2 flow, it depends also on the sticking coefficient (i.e. the probability of the N atom reacting with the Ta on the surface) which decreases as the N content increases. Thus, the N content in the film depends on the geometry of the sputtering system since the nitrogen is consumed by the Ta as it enters the deposition area. Also, the ions sputtering the target change from Ar^+ to N_2^+ as the N_2 flow increases. The N_2^+ ions provide N for reaction with the Ta target and the N atoms then have to be sputtered as well as the Ta; thus, the sputtering rate of Ta decreases as the N_2 flow increases. Oxygen is even more reactive and strong doping gradients result [29].

A5.0.3 GETTERING AND HYSTERESIS

Gettering reactions are evident when the reactive gas flow rate is increased while the metal target is sputtered in Ar. In the absence of sputtering, the pressure (p) increases with flow but, with sputtering, the total pressure does not increase until a critical flow is reached; then p increases rapidly to a higher value (figure A3.0.17). When the reactive gas flow is then varied, p increases or decreases until the flow is reduced to a much lower value when p decreases rapidly to the original value for sputtering in Ar. This variation in p is called the hysteresis for reactive sputtering. This term appears to have been first used by Geraghty and Donaghey [30] in diode sputtering of Ti in Ar/O_2. With the sputter rates available, the oxide could be removed only by reducing the O_2 flow to zero, to prevent replacement of the oxygen on the target.

With the higher sputtering rates in magnetrons, the pressure hysteresis was observed. Another hysteresis is observed in the target voltage required for constant current; V_T typically decreases as p increases rapidly and then returns to the original, higher value when p returns to its lower value.

For Al as an example, as $f(O_2)$ increases from zero, the oxygen is gettered by reaction with the sputtered Al until the arrival rate at the target exceeds the removal rate, which occurs first where the sputtering rate is lowest. Al_2O_3 forms on the target surface. As soon as some oxide forms, the sputtering rate of Al decreases due to the lower sputter yield for the oxide; this decreases the gettering rate and increases the oxygen available to oxidize the target, causing a rapid change from an Al target to an Al_2O_3 target surface. Due to the lower sputter rate of Al from this surface, less oxygen is gettered by the sputtered Al. Thus, $f(O_2)$ must be decreased substantially before there is an insufficient supply to maintain the oxide on the target surface.

The pressure hysteresis is due to gettering (i.e. pumping) of the reactive gas by the sputtered metal. This getter pump competes with the normal system pump to remove the reactive gas. If the speed of the latter pump is increased [31], the getter pump is less significant and p is essentially determined by the vacuum pump, so that there is only a very small change in the system pressure due to compound formation on the target; the greater the ratio of pump speeds, the smaller the hysteresis. However, the deposition rate and the film composition still change in the same way even if there is no observable pressure hysteresis. However, higher gas flow rates are required because of the higher pumping speed.

The decrease in voltage when the oxide forms on the target is probably due to the higher secondary electron yield from the oxide than from the Al [32]. The increase in electron yield increases the ionization rate so that the ions require less energy to produce sufficient electrons to maintain the discharge. Targets of different materials may give either a positive (e.g. Nb, Sn in O_2) or negative (e.g. Al, Si, Ti in O_2) change in V_T as the compound forms [33] even though one would expect the secondary electron coefficients of the oxides to be higher in all cases. Schiller *et al* [34] reported an increase in V_T before the decrease due to target oxidation; they attributed this to constriction of the current in the DC magnetron due to oxide formation on the target. Noel *et al* [35] found that V_T increased when TiN formed on a Ti target.

There is also a decrease in the deposition rate due to the lower sputtering rate for the compound layer on the target surface. Vireton *et al* [36] modelled the sputtering rate and estimated the thickness of the layer on the target surface as 1.5 nm. However, estimates from the time required to remove the layer from the target often give higher values, up to 100 nm. de Jesus *et al* [37] found that the formation of NiO on a Ni surface was accelerated by simultaneous bombardment by Ar^+; thus, the actual layer which is sputtered from targets may change with the ion species at the target.

Modelling of processes and the hysteresis during reactive sputtering is discussed in A5.3.

A5.0.4 SPUTTERED SPECIES

When a compound target (e.g. NiO) is sputtered by Ar^+ ions, the situation is the same as sputtering an alloy since a compound and an alloy are essentially the same as for as the sputtering mechanisms; i.e. a homoegeneous structure of atoms of different elements bonded together. The surface composition changes until a steady state when the sputtered flux represents the composition of the compound. In most cases, the anion (e.g. O) will be sputtered preferentially [38]. In the NiO case, the sputtered flux ratio is $n(O)/n(Ni)= 1$; if both O and Ni had the same sticking coefficient on the surface, the film would be NiO. However, O has a lower sticking coefficient and the films are slightly deficient in oxygen. For O to have a sticking coefficient of 1, the O atom would have to react with a Ni atom as it arrives at the substrate; otherwise, it is desorbed from the substrate.

To compensate for the oxygen deficiency, O_2 is added to the system. Now O atoms are added to the surface of the target, enriching the surface relative to the equilibrium composition in the absence of O_2. Thus, the $n(O)/n(Ni)> 1$ in the sputtered flux and this provides additional oxygen to the film. Thus, to obtain fully reacted films, O_2 must be added to the Ar whether the target is NiO or Ni.

Both atoms (Ni, O) and clusters (NiO) are sputtered from the NiO target. Coburn *et al* [39] measured the fraction of species like NiO in the sputtered flux from oxide targets using a mass spectrometer. The fraction increased with the metal–oxygen bond strength and with the additional oxygen supply. The probability of NiO being sputtered increases as the O surface coverage increases since more O atoms are sputtered and can combine with Ni as they are ejected together; the probability of combining also increases as the bond strength increases [40]. Oechsner *et al* [41] measured the yields of different species sputtered from Ta_2O_5 by post-ionization of the sputtered neutrals; TaO was the most abundant species. Stirling and Westwood [42], using an atomic absorption technique, found that there was no detectable Al flux sputtered from an oxidized Al target. These oxide results indicate that a large fraction of the sputtered species are in the form of clusters.

However, Petrov *et al* [43] concluded that there were few Ti-N clusters from a nitrided Ti target, even though the Ti-N binding energy is high, but Kumar *et al* [44] detected no Al atoms when depositing AlN. Poitevin and Lemperier [45] suggested that TiN was dissociated after ejection from the target in DC diode sputtering. Thus, there appears to be a range of results on the probability of clusters, depending on the system.

During reactive sputtering of oxides, it is necessary to consider the sputtering of O^- [46] and the effect of these highly energetic species on the deposition.

A5.0.5 ENERGY AT SUBSTRATE

These energetic sputtered species arriving at the substrate cause heating of the substrate and may also significantly affect the film properties. Higher energy and substate temperature usually densifies the film. While sputtered atoms are the principal component of the energetic flux, reflected atoms and negative sputtered ions will have much higher energies; even if they comprise only 10% of the flux, they may dominate the energy input.

The energy of the flux, obtained from calorimetric measurements [47], was as high as 1000 eV per deposited 'molecule' for Al_2O_3; even higher values were observed for AlN [48], probably due to the reflected N from the target. Hantzpergue *et al* [49] concluded that, when N_2^+ bombarded an Al target, one of the atoms reflected with half the incident energy. Petrov *et al* [43] proposed a similar effect in the case of TiN. However, TRIM calculations indicate that, when 240 eV N_2^+ sputters the TiN target, the N reflection coefficient is 12% and the average energy is only 22 eV [50].

In reactive sputtering, the chemical reaction energy may also be important. Poulek *et al* [51] suggested that when the target was nitrided TiN clusters were sputtered and the Ti-N reaction energy was not available at the substrate. Lee *et al* [52] found that the growth rate of AlN films decreased when H_2 was added to the Ar/N_2 due to the reaction of H_2 with the AlN being deposited; the orientation of the films changed from (002) to a mixed (100)/(110) orientation due to removal of the N from the surface which prevents the AlN growing with the *c*-axis normal to the surface. Aida and Rahmane [53] used a piezoelectric detector to measure the force of energetic particles; this would distinguish between heating and chemical energy contributions.

Of course, the energy at the substrate can be increased by biasing the substrate, either directly or by 'unbalancing' the magnetron. Wiemer *et al* [54] used Langmuir probe methods to estimate the energy delivered to the film by N_2^+ when TiN was deposited; the energy increased with bias to > 2000 eV/atom at −200 V substrate bias.

Hultman *et al* [55] contrasted films grown with a high flux of low energy (~ 20 eV) N_2^+ and a low flux of high energy ions. Increasing the flux, increased the film density and changed the orientation from (111) to (002). Films grown with high substrate bias at low ion flux tend to grow towards the most open channeling direction with defects introduced due to the collision cascades in the film. However, Lousa and Gimono [56] obtained h-BN at 80 eV but c-BN at 120 eV when the N_2^+ flux remained constant.

Reviews of the effects of ion bombardment on film properties can be found in [57, 58].

A5.0.6 OTHER METHODS

Arc prevention and AC sputtering techniques either prevent the arc or avoid the charge build up on targets and are becoming the main methods for reactive sputtering at high rates. The special characteristics of power supplies required to control arcs have been described by Scholl [59]. The methods are covered in sections A5.1 and A5.2. However, other approaches have been used and some are summarized here.

Baffled magnetron

In the baffled magnetron [60], Ar is injected into an enclosure round the target while O_2 is injected close to the substrate. The target and substrate are separated by a baffle plate to reduce the oxygen flux to the target. The baffle separates the discharge from the substrate, which is a disadvantage for reaction at the substrate. A low power RF discharge, or another method for increasing the reactivity, is established at the substrate to promote reaction without significant resputtering. The baffle is designed to reduce the conductance of O_2 to the target while minimizing the deposition of Al on it since this reduces deposition on the substrate. However, Al deposited on the walls of the apertures in the baffle reduces the oxygen flux to the target by gettering it. The process tolerance depends on the baffle design; the more reactive the metal and gas, the more restrictive the baffle must be; at too high a flow, oxide forms on the target and arcing of the DC supply occurs.

In some cases, such as indium tin oxide, a single large aperture and a fast moving substrate have been used [61]. An active, rather than a passive, baffle was used by Belkind and Wolfe [62] to deposit TiO_2; two rotating cylinder planar magnetrons provided the baffle action while Ar was introduced into the planar magnetron target region. The rotating targets were operated in the oxide mode while the planar was in the metal mode. Oxygen passing the baffles reforms the oxide sputtered from the rotating magneterons.

Separate deposition and oxidation

In the separate chamber systems [63], the substrate is rotated between two chambers so that a thin metal layer is deposited in one, by sputtering the metal in Ar, and then oxidized in the other and this is repeated till the film deposition is complete; the rotation rate must be fast enough that the deposited layer can be fully oxidized; the metal layer must be only a few monolayers thick. An RF discharge sustained at the substrate holder increases the reaction rate. In the 'Metamode' version of the separate chamber system [64], a special magnetron ion source accelerates O_2^+ across the anode sheath to the moving substrate; the ion current densities were 3–6 mA cm^{-2}. The substrate holder drum rotates at 80 rpm. Al_2O_3 films with refractive index of 1.65 were deposited at 10 nm s^{-1} at 2 mTorr.

Another method for providing the activated oxygen for the reaction is to use an ECR plasma generator to inject oxygen ions and atoms into the region in front of the substrate [65]. Ions from ECR plasmas typically have energies of 5–30 eV. In A3.2, Penfold discussed the coupling of the ECR and magnetron fields for efficient sputtering of Al_2O_3 [66].

Increasing the distance from target to substrate also separates the sputtering and deposition processes; the large distance reduces the possibility of the reactive gas which is introduced near the substrate reaching the target, especially if high gas pumping speed is maintained. However, since the magnetron plasma is far from the substrate, the oxidation requires delivery of reactive gas ions or some other active species to the substrate [67]. This approach is similar to ion beam reactive deposition, as described by Harper *et al* [68], where an ion beam delivers N_2^+ to the substrate to react with Al from a target sputtered by an Ar^+ beam; the N_2^+ cause some resputtering of the Al.

Gas flow modulated sputtering

Sekiguchi *et al* [69] used O_2 pulses while sputtering a Ti target. They found that the deposition rate increased slightly with increasing period of the pulse; in each period the O_2 was on for half the period. The O/Ti ratio in the film was roughly constant until the period was 20 s, when the films were metallic and the rate increased by an order of magnitude. This was prevented by higher O_2 flow rates during the pulse; i.e. more oxidation.

Chemical reaction sputtering

To prevent oxidation of a target in a suitable case, the target could be run hot to evaporate the oxide as it formed; however, this would have limited application and the substrate would be heated. A more sophisiticated way of accomplishing this for Si [70] adds another gas (Cl_2) to the system to etch the oxide formed on the metal target; the etching reaction occurs due to the ion bombardment of the target. The resulting oxide free Si target is sputtered and SiO_2 is deposited as long as there is no atomic Cl at the substrate. However, this is obviously of limited application since it depends on having the appropriate chemistry at both target and film surface.

Reactive gas flow sputtering

Jung and Westphal [71] used a very high Ar gas flow to sweep sputtered Al out of a linear hollow cathode to the substrate with O_2 injected in front of the substrate. Al_2O_3 was deposited at 1.8 nm s^{-1} kW^{-1}, but the absorption was high so the reaction was incomplete.

A5.0.7 SUMMARY

The properties of the film deposited by reactive sputtering have not been considered since they are best dealt with specifically by material elsewhere in the Handbook. The large number of parameters which vary from system to system make comparisons difficult and the understanding of the processes is very incomplete.

REFERENCES

[1] Grove W R 1852 *Phil. Trans. R. Soc.* **142** 87
[2] Berry R W 1961 US Patent 2,993,266
[3] Westwood W D, Waterhouse N and Wilcox P S 1975 *Tantalum Thin Films* (New York: Academic)
[4] Westwood W D 1989 *Physics of Thin Films* **14** 1
[5] Westwood W D 1990 *Handbook of Plasma Processing Technology* ch 9 (Noyes Publications)
[6] Lewis D C, Westwood W D and Sadler A G 1964 *J. Can. Ceram. Soc.* **33** 153
[7] Cheng Y C and Westwood W D 1974 *J. Electronic Mater.* **3** 37
[8] Davidse P D and Maissel L I 1966 *J. Appl. Phys.* **37** 574
[9] Minakata M, Chubachi N and Kichuchi Y 1972 *Jpn. J. Appl. Phys.* **11** 1852
[10] Este G O and Westwood W D 1988 *J. Vac. Sci. Technol.* A **6** 1845
[11] Swann S 1988 *Vacuum* **38** 791
[12] Gnaedinger R J 1969 *J. Vac. Sci. Technol.* **6** 355
[13] Sproule W D, Wong M S, Lopez S, Li D and Scholl R A 1995 *J. Vac. Sci. Technol.* A **13** 1188
[14] Frach P, Heisig U, Gottfried C and Walde H 1993 *Surf. and Coatings Technol.* **59** 177
[15] Lessard P A 1990 *J. Vac. Sci. Technol.* A **8** 2874

[16] Howson R P 1994 *Pure Appl. Chem.* **66** 1311
[17] Taga Y 1993 *Appl. Opt.* **32** 5519
[18] Sproul W D 1994 *J. Vac. Sci. Technol.* A **12** 1595
[19] Mattox D M 1996 *Surf. and Coatings Technol.* **81** 8
[20] Hara T, Kitamura T, Tanaka M, Kobayashi T, Sakiyama K, Onishi S, Ishihara K, Kudo J, Kino Y and Yamashita N 1996 *J. Electrochem Soc.* **143** L264
[21] Stavrev M, Fischer D, PreuB A, Wenzel C and Mattern N 1997 *Microelectron. Eng.* **33** 269
[22] Shimooka Y, Iijima T, Nakamura S and Suguro K 1997 *Jpn. J. Appl. Phys.* **36** 1589
[23] Auciello O, Krauss A R and Gifford K D 1994 *Ferroelectric Thin Films Synthesis and Basic Properties* (New York: Gordon and Breach)
[24] Martin P J and Netterfield R P 1995 *Handbook of Ion Beam Processing Technology* ch 19 (Noyes Publications)
[25] Dobrowolski J A, Pekelsky J R, Pelletier R, Ranger M, Sullivan B T and Waldorf A J 1992 *Appl. Opt.* **31** 3784
[26] Gregory O J, Slot A B, Amons P S and Crisman E E 1996 *Surf. and Coatings Technol.* **88** 79
[27] Blanton T N and Batterjee D K 1995 *Thin Solid Films* **256** 59
[28] Wasa K and Hayakawa S 1992 *Handbook of Sputter Deposition Technology* (Noyes Publications)
[29] van Zon J B A D and Rijpers J C N 1989 *J. Vac. Sci. Technol.* A **7** 144
[30] Geraghty K G and Donaghey L F 1976 *J. Electrochem Soc.* **123** 1201
[31] Okamoto A and Serikawa T 1986 *Thin Solid Films* **137** 143
[32] Szapiro B and Rocca J J 1989 *J. Appl. Phys.* **65** 3713
[33] Munz W D, Heimbach J and Reineck S R 1981 *Thin Solid Films* **86** 175
[34] Schiller S, Heisig U, Steinfelder K, Strumpfel K and Seiber W 1980 *Vakuum-Teknik* **30** 3
[35] Noel J P, Houghton D C, Este G, Shepherd F R and Plattner H 1984 *J. Vac. Sci. Technol.* A **2** 284
[36] Vireton *et al* 1995 *Nucl. Instrum Methods* B **95** 34
[37] de Jesus J C, Pereira P, Carrazza J and Zaera F 1996 *Surf. Sci.* **369** 217
[38] Kelly R 1989 *Handbook of Plasma Processing Technology* ch 4 (Noyes Publications)
[39] Coburn J W, Taglauer E and Kay E 1974 *Jpn. J. Appl. Phys.* **S2-1** 501
[40] Hecq A, Vandy M and Hecq M 1980 *J. Chem. Phys.* **72** 2876
[41] Oechsner H, Schoof H and Stumpe E 1978 *Surf. Sci.* **76** 343
[42] Stirling A J and Westwood W D 1971 *Thin Solid Films* **7** 1
[43] Petrov I, Myers A, Greene J E and Abelson J R 1994 *J. Vac. Sci. Technol.* A **12** 2846
[44] Kumar N, Pourrezaei K, de Maria R J and Dingh B 1986 *Mater. Res. Symp. Proc.* **68** 357
[45] Poitevin J M and Lemperier G 1984 *Thin Solid Films* **120** 223
[46] Tominaga K, Chong M and Shintani Y 1994 *J. Vac. Sci. Technol.* A **12** 1435
[47] Thornton J A and Lamb J L 1984 *Thin Solid Films* **119** 87
[48] Este G O and Westwood W D 1987 *J. Vac. Sci. Technol.* A **5** 1892
[49] Hantzpergue J J, Pauleau Y, Bouteville A and Remy J C 1985 *Le Vide* **225** 13
[50] Biersack J P and Haggmark L G 1980 *Nucl. Instrum. Methods* **174** 257
[51] Poulek V, Quaeyhaegens C, Knyut G, Stals L and Fagard V 1993 *Surf. and Coatings Technol.* **60** 480
[52] Lee H-C, Lee K-Y, Yong Y-J, Lee J-Y and Kim G-H 1995 *Thin Solid Films* **271** 50
[53] Aida M S and Rahmane S 1996 *Thin Solid Films* **288** 83
[54] Wiemer C, Levy F and Messier R 1996 *J. Phys. D: Appl. Phys.* **29** 99
[55] Hultman L, Sundgren J-E, Greene J E, Bergstrom D B and Petrov I 1995 *J. Appl. Phys.* **78** 5398
[56] Lousa A and Gimono S 1997 *J. Vac. Sci. Technol.* A **15** 62
[57] Cuomo J J, Rossnagel S M and Kaufman H R (eds) 1989 *Handbook of Ion Beam Processing Technology* (Noyes Publications)
[58] Colligon J S 1995 *J. Vac. Sci. Technol.* A **13** 1649
[59] Scholl R A 1997 *Surf. and Coatings Technol.* **93** 7
[60] Maniv S, Miner C J and Westwood W D 1981 *J. Vac. Sci. Technol.* **18** 195
[61] Howson R P and Ridge M I 1981 *Thin Solid Films* **77** 119
[62] Belkind A and Wolfe J 1994 *Thin Solid Films* **248** 163
[63] Schiller S, Heisig U, Beister G, Steinfelder K, Strumpfel J, Korndorfer C and Sieber W 1984 *Thin Solid Films* **118** 255
[64] Lehan J P, Sargent R B and Klinger R E 1992 *J. Vac. Sci. Technol.* A **10** 3401
[65] Fujii T, Anno T, Koyanagi T, Hirai H and Matsubara K 1991 *Jpn. J. Appl. Phys.* **30** 1248
[66] Takahashi C, Kuichi M, Ono T and Matsuo S 1988 *J. Vac. Sci. Technol.* A **6** 2348
[67] Scobey M A 1996 US Patent 5,525,199
[68] Harper J M E, Cuomo J J and Hentzell H T G 1983 *Appl. Phys. Lett.* **43** 547
[69] Sekiguchi H, Kanzawa A, Imai T and Honda T 1994 *J. Vac. Sci. Technol.* A **12** 3176
[70] Nandra S S 1990 *J. Vac. Sci. Technol.* A **8** 3179
[71] Jung T and Westphal A 1993 *Surf. and Coatings Technol.* **59** 171

A5.1 Reactive Pulsed DC Magnetron Sputtering and Control

Jochen M Schneider and William D Sproul

A5.1.0 INTRODUCTION

This section presents a discussion of reactive pulsed DC magnetron sputtering and control. Since sputtering and magnetron sputtering effects are discussed in depth in chapters A3.0 and A3.2, respectively, they are considered here only as required in discussing the reactive pulsed DC magnetron sputtering process.

Reactive sputtering deals with sputtering in a reactive gas or a mixture of reactive gas and inert gas atmosphere. The formation of a compound, the reaction product of the sputtered flux and the reactive gas, is desired.

Today, many complex compounds [1] and multilayered designs [2] can be realized with reactive magnetron sputtering. Deposition rates close to that of the pure sputtered material can be achieved for both conductive [3] and non-conductive materials [4] and the combination thereof [5].

The bulk of this chapter is divided into three sections. After briefly reviewing the basics of the reactive magnetron sputtering process in A5.1.1, much of this section deals with recent advances in reactive sputtering. The enabling technological tools for sputtering non-conductive materials at high deposition rate are medium frequency pulsed DC power, as discussed in A5.1.2, in combination with controlled compound formation at the target surface, which is discussed in A5.1.3. Both key technologies are critically reviewed, and the state of the art is presented.

A5.1.1 BASICS OF REACTIVE MAGNETRON SPUTTERING

A5.1.1.1 Effect of the reactive gas on the target surface

In reactive magnetron sputtering, the reactive gas can form a compound layer on the target surface. The compound formed as a reaction product of the target material with the reactive gas usually has a different secondary electron emission coefficient from the pure target material that results in a change of the plasma impedance and the target potential if the target power is kept constant. The target current constitutes the sum of the ion current impacting on the target and the electron current. Hence, the ionic component of the current is changed as the electron current changes. The formation of a compound layer on the target surface such as Al_2O_3 results in a lower deposition rate for several reasons. The sputter yield for alumina is lower than for aluminum, and the secondary electron emission coefficient for the oxide is higher than that of the metal. The plasma impedance drops with the injection of the extra secondary electrons, and ions that bombard the surface do so with less energy which reduces the sputtered flux, and hence the net deposition rate, even further.

In the past, researchers distinguished between two different sputter modes [6]: firstly, sputtering of a metallic target at low reactive gas partial pressures and, secondly, sputtering from the compound target (or poisoned target), at large reactive gas partial pressures, as described at the beginning of this section. It was believed that at a certain transition point [7, 8] the target changes from the metallic sputtering mode to the poisoned mode. Heller [9], Gornachev *et al* [10, 11] and Geraghty and Donaghey [12] showed that the sputtering mode can be identified from the target current/voltage characteristic. Schiller *et al* [13] showed that the transition point can in fact be a transition region, which in the past was frequently referred to as the 'unstable region', due to the fact that stable long term operation was not possible in this region. However it was quickly realized that this region is of high technological significance if the deposition of stoichiometric compounds at high rate is desired. The efforts that have been made to access this region and deposit coatings at high rate with a defined chemical composition are discussed in detail in section A5.1.3. Sproul *et al* demonstrated that, by employing an automatic feedback control technique, the deposition of transition metal-nitrides [14, 15] and carbides [3] could be carried out in the 'unstable' region. For example, the formation of stoichiometric TiN was reported at 100%, and HfN and ZrN at 95% of the metal deposition rate [14]. High rate deposition with precise control of the coatings stoichiometry was achieved by maintaining a certain partial pressure value of the reactive gas precisely throughout the deposition process. At constant reactive gas partial pressure, the sputtered flux is kept constant automatically, and the chemical composition and film homogeneity are maintained, since the partial pressure ratio of Ti to N_2 is constant. See section A5.1.3 for a more detailed discussion.

A5.1.2 Reactive Magnetron Sputtering of Non-Conductive Materials—Pulsed DC Magnetron Sputtering

The reactive gas partial pressure required for the formation of a particular stoichiometric compound at the substrate may result in a partially reacted target surface. Stable operation of the sputter source can be maintained if the formed compound is a conductor. In the case of a non-conductive compound layer, such as TiO_2, AlN or Al_2O_3, the positive ions impinging on the target cannot be neutralized by electrons from the target, which leads to accumulation of positive charge on the target surface.

Electrically this situation represents a capacitor. If the dielectric strength of the compound is larger than the potential difference between the dark space side and target side of the compound layer, more and more charge is accumulated. At the same time, the voltage drop across the cathode dark space is reduced to the point where the impinging ions are too low in energy to cause sputtering. If the dielectric strength of the compound layer is smaller than the potential difference between the dark space side and the target side of the insulating layer, the compound layer will break down. This situation leads to the concentration of a large part of the discharge current in a small surface volume segment and causes local evaporation. This process is known as 'arc evaporation' [16] of the target material, and is accompanied by two undesirable effects: the ejection of macroparticles, leading to inhomogeneity and defects in the film, and a drastic change in processing parameters, such as cathode potential and the vapor to reactive gas concentration ratio. Arcing represents a problem for the control of the sputtering process and generally leads to instabilities.

Until the advent of AC sputtering or medium frequency pulsed DC sputtering, insulating films such as alumina or zirconia were mainly deposited by RF sputtering. Since the sputter yield of, for example, alumina is lower than the yield for aluminum, it is well known that the sputtered flux using RF sputtering is at least a factor of 10 smaller than by DC sputtering [17].

The first report of AC sputtering at a frequency of 60 Hz was for metal and alloy films, and this work was done in 1972 by Kumagai [18], who reported on the AC sputtering of Ta and Ta alloy films. The pioneering work for AC or pulsed DC sputtering of non-conductive compounds was carried out by Cormia *et al* at Airco 20 years ago [19]. Estimating the thickness of the dielectric compound layer to be on the order of several μm, he calculated the required frequency to avoid arcing to be of the order of several kHz. Utilizing the electron current to discharge the positive charge accumulated at the compound surface, arc free, reactive magnetron sputtering of TiO_2 was possible at a frequency of 10 kHz. By utilizing a conventional anode to create a discharge for the time that the target potential is positive, a large positive voltage is required at the target to maintain the diode mode. Simultaneous sputtering of the anode occurs.

The process described in Cormia's patent actually splits into two sub-processes. Sputtering occurs on the negative half-cycle, and excess positive charge is neutralized by electrons on the positive half-cycle.

In 1986 and 1988, patents were issued in both the former East [20] and West Germany [21] for pulsed power in the kHz frequency range. In the same time period, Este and Westwood [22] published their work on rate enhanced sputtering of dielectric materials by a 'quasi-direct-current sputtering technique'. The influence of the frequency on the deposition rate was investigated for AlN films. The individual deposition rates were normalized to the deposition rate of the DC case and are given as a function of frequency in figure A5.1.1. Este and Westwood found that the RF deposition rate was half of the DC rate, and in the kHz range the deposition rate was reduced by 10% to 25% of the DC rate. The film properties were not affected by the frequency. The frequency range of today's commercial power supplies that will suppress the electrical breakdown of insulators (approximately 10^8 V m^{-1}) at normalized deposition rate losses of less than 20% is shown in figure A5.1.1. These losses can be reduced if the duty cycle can be increased.

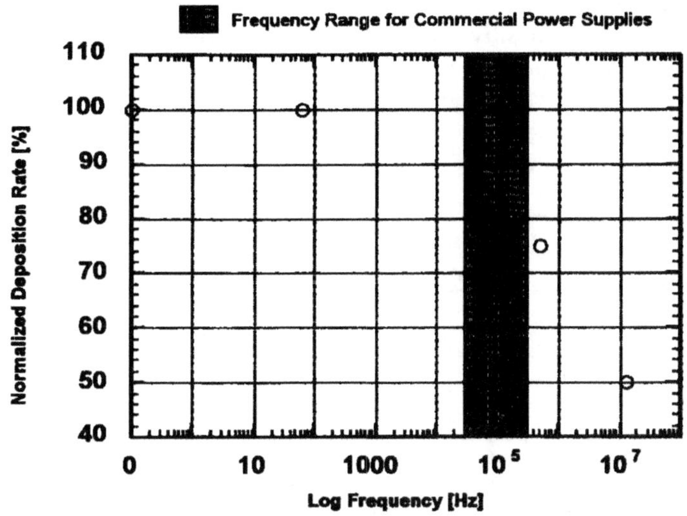

Figure A5.1.1. Normalized deposition rate versus frequency, after Este and Westwood [22].

In 1991 Scherer *et al* deposited Al_2O_3, SiO_2 and Si_3N_4 by reactive AC magnetron sputtering [23]. The deposition technique is based on two magnetron sources. The voltage output of the AC power supply driving the sources is supplied with a phase shift of 180°. In this way the two sputter sources are run periodically as a cathode and an anode.

Today many commercial pulsed DC power supplies with frequencies of up to 200 kHz are available. These pulsed DC power supplies produce either a bipolar symmetric or asymmetric pulse as are shown schematically in figure A5.1.2. With the symmetric pulsed DC power, the pulse height is of equal magnitude for both the positive and negative pulses. In between the pulses there is an off time, and the width of positive pulse is usually smaller than the width of the negative pulse. The idea is to spend just enough time to discharge the target surface with the positive pulse, but since no sputtering takes place during the positive pulse, it should be as small as possible to get the job done.

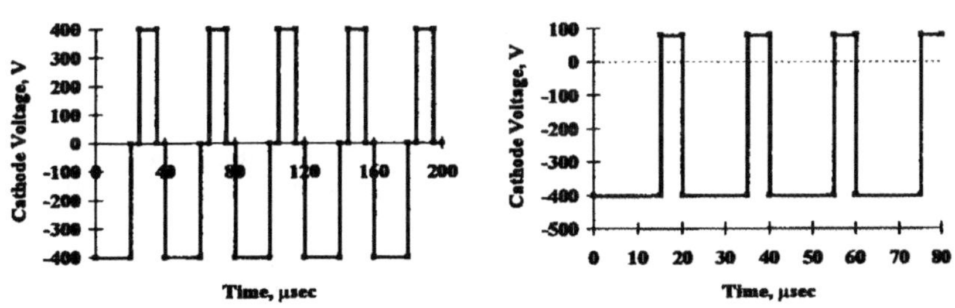

Figure A5.1.2. (*a*) Schematic drawings of bipolar (*a*) symmetric and (*b*) asymmetric pulsed DC power, after Sellers [55].

For the asymmetric pulsed DC power supplies, the magnitude of the positive pulse is only a fraction of the magnitude of the negative pulse, and the sense of the pulse changes directly from positive to negative with no off time. Similar to the symmetric bipolar pulsed power, the width of the positive pulse is usually smaller than the negative pulse. With both the bipolar symmetric and asymmetric pulsed DC power supplies, sputtering takes place from the target during the negative pulse, whereas discharging of the target surface takes place during the positive pulse.

Commercially there is another way of achieving bipolar asymmetric pulsed DC power. Instead of producing a pulsed output directly from the power supply, a normal DC power supply is used, but a switching network is placed between the DC power supply and the sputtering target. This switching network produces a pulsed DC output from a conventional DC input.

Both ways of producing bipolar asymmetric pulsed DC power work well, but there can be differences in the actual pulse shape at the target. A typical pulse pattern from the switched asymmetric pulsed DC supply is shown in figure A5.1.3 [24]. It looks very close to the ideal schematic drawing. There is little overshoot with either the positive or negative pulses. However, for the asymmetric pulsed DC power delivered directly from the pulsed DC power supply, there can be significant overshoot particularly with the negative pulse as is shown in figure A5.1.4. This overshoot is a function of the way that the power supply operates. The average target voltage for this example was about −450 V, but the actual peak-to-peak value was almost 1500 V.

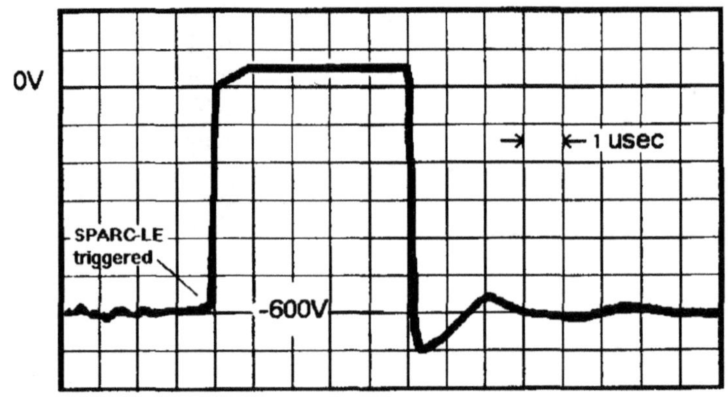

Figure A5.1.3. Voltage versus time trace for asymmetric pulsed DC power from a switched source [56].

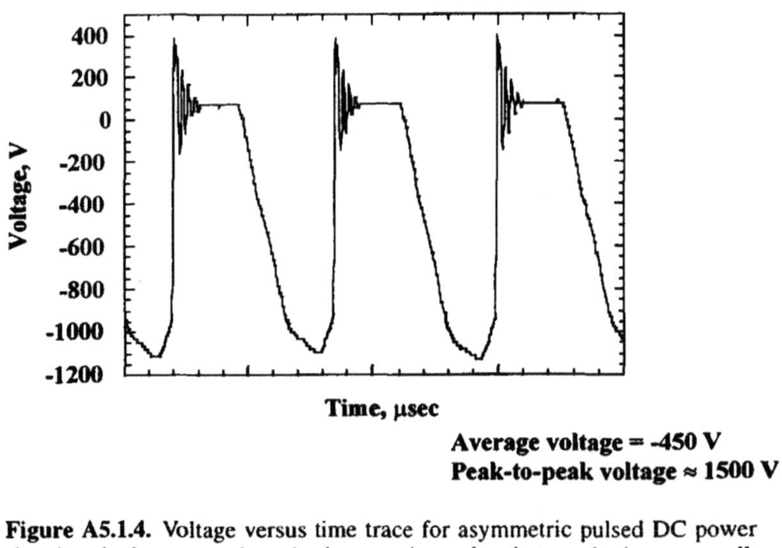

Figure A5.1.4. Voltage versus time trace for asymmetric pulsed DC power showing the large overshoot in the negative pulse that results in an overall peak-to-peak voltage of about 1500 V.

The voltage overshoot that occurs from this pulsed DC power supply is due to the fact that the power supply is actually two power supplies in one housing. A constant current power supply is used for the negative pulse whereas a constant voltage supply is used for the positive pulse. The large voltage overshoot occurs in the negative pulse when the constant current power supply initially turns on and is trying to reach the current operating set-point.

The effects of this large overshoot in the negative pulse are still being determined, but it is safe to say that the average energy of the high energy particles is larger in the direct pulsed DC discharge, than in the switched DC discharge. Due to the rather large peak currents, the plasma density is probably increased in the direct pulsed DC discharge.

If pulsed DC power is used to bias a magnetron source and/or the substrate, it is important to acknowledge that both the ion and neutral energies, as well as the ion and neutral fluxes impinging on the growing film, are a function of time. If more than one pulsed DC power supply is employed to drive a discharge, it is recommended to synchronize them in a master–slave set-up. For example, if pulsed DC power is applied to both the target and the substrate, both power supplies should pulse positively at the same time with the same frequency, and both should pulse negatively together. This can be accomplished by having one of the pulsing units act as the master, and the other (slave) mimicking the pulse pattern of the master. Such master–slave set-ups are available commercially.

A5.1.3 PROCESS CONTROL OF REACTIVE MAGNETRON SPUTTERING

A5.1.3.1 Why is process control needed in reactive magnetron sputtering?

In subsection A5.1.1, it was discussed that in reactive magnetron sputtering the reduction in deposition rate is due to the formation of a compound layer on the target surface and the subsequent reduced sputtering yield. For any industrial application, the deposition rate or rather the maximization of the deposition rate, is of major importance. Not only the deposition rate but also the film properties show a strong dependence on the ratio of sputtered flux to the reactive gas concentration. For example, the refractive index of indium tin oxide (ITO) [25] or the hardness and preferred orientation of TiN [26] is a direct function of the partial pressure of the reactive gas in the sputtering chamber during the deposition of the film. The motivation for the development of reactive sputtering controllers is obvious: firstly to control and in some cases maximize the deposition rate, and secondly to control the film stoichiometry, and hence microstructure and properties.

A5.1.3.2 How can the reactive sputtering process be controlled?

Early works concerned with the mechanism of reactive sputtering, especially with the step-like decrease of the deposition rate of metals sputtered in a reactive gas atmosphere [7, 8, 27, 28] and with basic investigations on the influence of a reactive gas on discharge current [9] and current–voltage characteristic [10, 11] as well as research into the kinetics of reactive sputtering [12, 29] were instrumental for the development of reactive sputtering controllers. Different control concepts have been suggested over the years; separation of the reactive gas from the target by Schiller *et al* in 1979 [13], pulsing of the reactive gas flow by Thornton in 1979 [30], automatic feedback control of the reactive gas flow, using light emitted by the plasma by Chapin and Condon in 1979 [31], the target voltage by Schiller in 1982 [32], the total pressure by Ridge and Howson in 1982 [25], and the partial pressure of the reactive gas by Sproul *et al* in 1983 [14, 15] as the control parameter. Also, light emitted by the plasma was used to control the power supplied to the cathode by Enjouji *et al* in 1983 [33].

It has been shown that if the pumping speed can be increased beyond a critical value [34–37] no high-speed controllers are necessary because the transition from the metal to compound sputter mode is a gradual transition. However, for high volume vacuum chambers, as employed for industrial applications, the pumping rate requirements are large, and therefore rather expensive. The advantages and disadvantages of these control concepts will now be discussed in detail, and their further development and relevance critically reviewed.

A5.1.3.3 The complete separation of reactive gas from the target

The complete separation of the reactive gas from the target, in such a way that no compound film is formed on the target, was first suggested by Schiller in 1979 [13]. This can be achieved for TiO_2 and Ta_2O_5 by using the target voltage characteristic [38] to estimate the critical oxygen partial pressure range where sputtering in the metallic mode occurs. For other non-conducting compounds, variations of the basic idea were successfully employed. To increase the reactive gas concentration at the substrate, while sputtering a metallic target, a slit or baffle arrangement can be placed between the source and the substrate. Maniv, Miner and Westwood in 1981 [39] employed slits for the deposition of Cd_2SnO_4. Este and Westwood in 1984 [40] and Scherer and Wirtz in 1984 [41] employed a collimator, whereas Pang *et al* [42] employed an aperture arrangement for the deposition of alumina in 1989. For all of these techniques, where mechanical grids and similar devices are utilized, 60% to 90% of the sputtered flux is deposited on the grid [17]. In 1989 Jones and Logan [17] utilized a hollow cathode substrate holder to activate the oxygen at the substrate, while sputtering off a metallic target, and produced stoichiometric alumina films. However this technology was developed for flat substrates, and it is difficult to utilize it for three-dimensional substrates.

A5.1.3.4 Pulsing the reactive gas flow

Pulsing the reactive gas flow as a method of control was proposed by Thornton in 1979 [30]. As the reactive gas flow is pulsed (switched on then switched off), the target periodically cycles through the metallic state to the compound state. This technology was further developed by Aronson in 1980 [43] and Sproul in 1984 [44]. Most recently Billard *et al* [45] modulated the power supplied to the sputter source, essentially achieving the same effect as pulsing the reactive gas. The target surface periodically cycles through the compound and metallic modes. While providing significant improvements in growth rate, 60% of the metal deposition rate for TiN was reported by Aronson [43], it is likely with all these slow pulsing techniques (frequencies of the order of 1 Hz), there will be a variation in the chemical composition of the film at the same frequency as the pulsing of the power or reactive gas flow takes place.

A5.1.3.5 Optical emission control

An optical spectrometer was employed to control the reactive gas flow by Chapin in 1979 [31]. The optical wavelength selected is sensitive to the reactive gas partial pressure, has a sufficient peak to background ratio, and has no peak overlap with other wavelengths. The concept was successfully employed by Schiller in 1987 [46] and Howson in 1989 [47] for the deposition of ITO and indium oxide, respectively. Enjouji 1983 [33] employed an optical spectrometer to control the power supplied to the sputter source and also demonstrated control of the rate and stoichiometry for the deposition of ITO.

A5.1.3.6 Target voltage and target current control

Schiller suggested in 1982 [32] using the target voltage to control the reactive gas flow. Successful application was demonstrated for Cd_2SnO_4 by Schiller [32] and for AlN by Affinito and Parsons in 1984 [48]. Interestingly, with the same set-up, Affinito and Parsons were not able to deposit AlO_x because their feedback loop was too slow for the oxide. The target voltage versus reactive gas partial pressure characteristic is a function of materials involved in the process. At constant target power, alumina, for example, shows one maximum and one minimum, whereas chromia shows two maxima and one minimum, as is shown in figures A5.1.5 and A5.1.6 respectively. Alumina is a good material for voltage control when the target power is held constant because a small oxygen partial pressure variation results in a large voltage change, and the voltage drops consistently as the oxygen partial pressure is increased until a minimum is reached. The changes in oxygen partial pressure are in the low 10^{-4} Torr range, and they have a small effect on the total system pressure, which is usually in the mid-mTorr range.

When reactively sputtering chromia, the target voltage first decreases and then, after reaching a minimum value, increases as the oxygen partial pressure is increased. If voltage control is utilized for this material, a positive and negative controller has to be utilized for the individual slope regions of the target voltage–oxygen partial pressure characteristic.

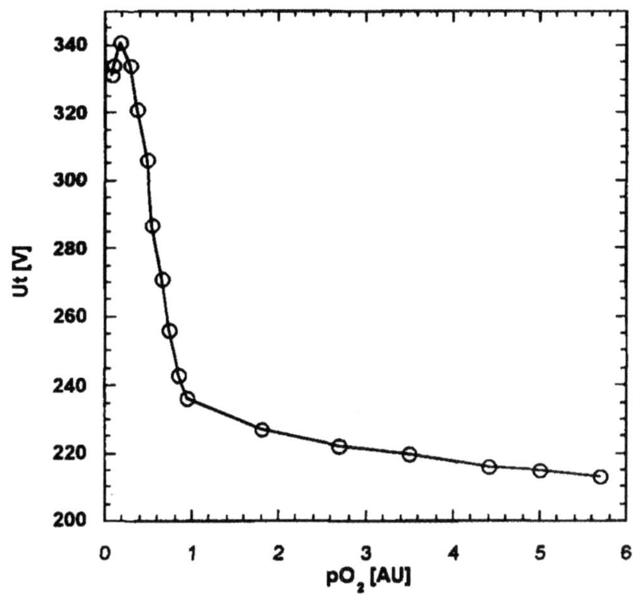

Figure A5.1.5. Alumina target voltage versus oxygen partial pressure [57].

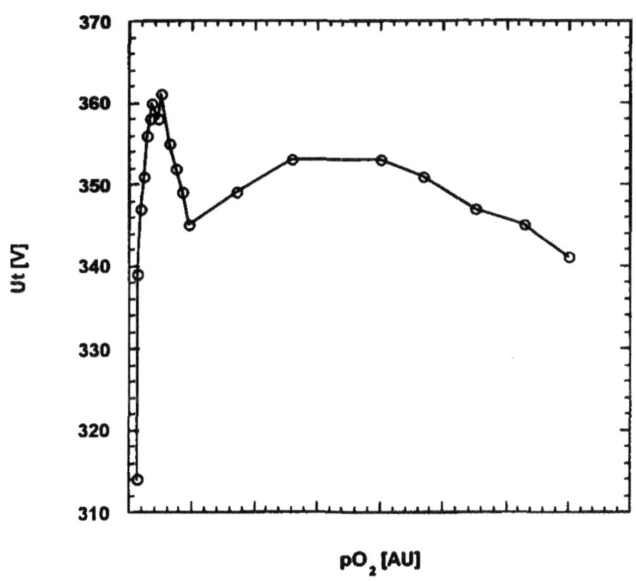

Figure A5.1.6. Chromia target voltage versus oxygen partial pressure [57].

A5.1.3.7 Reactive gas partial pressure control

Sproul and Tomashek [14, 15] introduced partial pressure control of the reactive gas in 1983. The reactive gas partial pressure is measured by a differentially pumped mass spectrometer, and the reactive gas flow is adjusted through automatic feedback control of the reactive gas to keep a constant reactive gas partial pressure. The ratio of the sputtered flux and the reactive gas partial pressure are kept constant, and hence the films are homogeneous. Very high rate deposition of the nitrides and carbides of Ti, Hf and Zr has been demonstrated by Sproul *et al* [3, 14, 26, 49, 50]. The partial pressure controlled hysteresis curve for the formation of TiN can be seen in figure A5.1.7.

Recently mass spectrometers that will operate at sputtering pressures have been developed. Although there is an obvious advantage at not having to differentially pump such a mass spectrometer, the sensitivity of

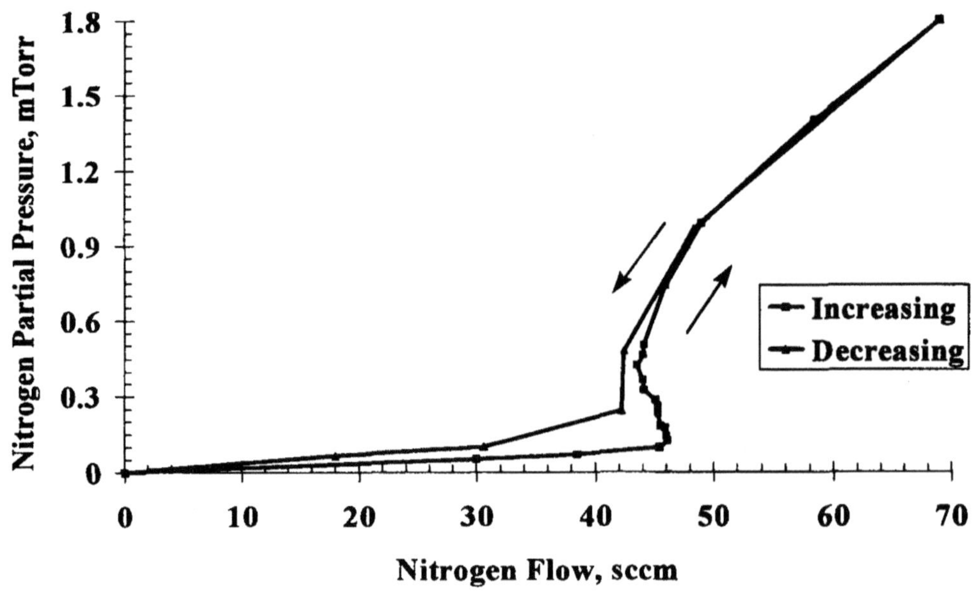

Figure A5.1.7. Partial pressure controlled hysteresis curve of TiN [58].

the device is not yet good enough to control the reactive sputtering process. Perhaps, in time, these devices will improve and be able to provide an accurate and rapid feedback signal for the control process.

For the nitride, the operating point on the hysteresis curve is usually at the nose of the curve where the partial pressure of the reactive gas just starts to rise. Typically for the nitrides of titanium, zirconium, and hafnium, this partial pressure is in the 0.1–0.3 mTorr range. Higher partial pressures can be used, but the deposition rate will be decreased. It may be advantageous to give up a little of the deposition rate if the nitride coating is being used for a decorative application for the color of the coating will vary from a light gold, to gold, to a brownish gold as the partial pressure is increased.

The obvious drawback of utilizing a differentially pumped mass spectrometer is that sampling has to take place in a remote area to the region of interest. The mass spectrometer measures the partial pressure of the reactive gas at the point where it is mounted on the system, not in front of the sputtering target. Rossnagel [51] and Sproul *et al* [50] studied the effects of gas heating and rarefaction directly in front of the target, and Sproul *et al* actually measured the reduction of the partial pressure of the reactive gas, which is reduced in front of the sputtering target due to energetic collisions with the sputtered atoms. However, from experience, the mass spectrometer provides a feedback signal to the controller that is proportional and repeatable.

Ridge and Howson [25] utilized an indirect pO_2 measurement via the total pressure at constant Ar flow rate to control flow of O_2 for the deposition of ITO.

The reactive DC magnetron sputter deposition of non-conducting oxides until just recently has almost been an impossible task. Reactive sputter deposition of the non-conducting oxides traditionally has been done at very slow rates using RF power. However, with the combination of pulsed DC power and partial pressure control of the reactive gas, it is now possible to reactively sputter oxides such as TiO_2 or Al_2O_3 at high deposition rates.

When flow control of the reactive gas is used for the reactive sputtering of a material such as TiO_2, initially all of the reactive gas will be consumed in the reaction with the metal being sputtered from the target as the reactive gas flow is increased from a low level (at a constant target power), as is shown in figure A5.1.8. When the gas flow reaches a certain level, there will be sufficient reactive gas to form a compound on the substrate, but this same compound will also form on the target surface. This reaction of the reactive gas with the target surface occurs very quickly for many oxides, and when the target becomes covered with the compound (poisoned), the sputtering rate drops rapidly. Due to the lower sputtering rate, less reactive gas is consumed, and its partial pressure jumps up. With this flow control of the reactive gas, there is a

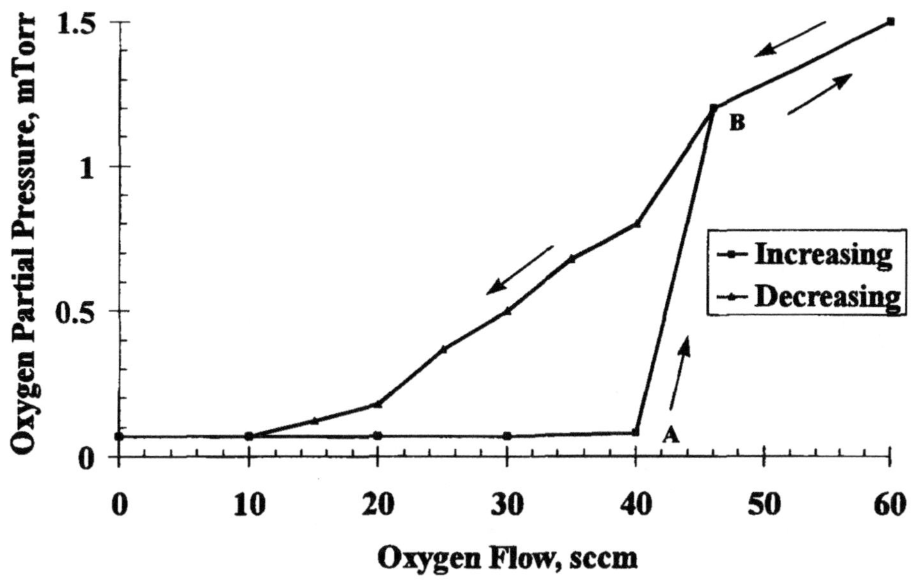

Figure A5.1.8. Hysteresis curve for the reactive pulsed DC magnetron sputtering of titanium in an argon/oxygen atmosphere using flow control of the reactive gas [59].

whole range of forbidden compositions between points A and B in figure A5.1.8 that cannot be deposited with this technique.

When the target is fully poisoned, any further increases in reactive gas partial pressure lead to a linear increase in the reactive gas partial pressure. When the gas flow is reduced, it takes time for the compound on the surface of the target to be removed by the sputtering process and for metal to be sputtered again. When the compound is broken through, the partial pressure of the reactive gas drops due to the reaction between the large flux of sputtered metal and the reactive gas. The drop in the partial pressure when the flow is reduced completes the hysteresis curve for this reactive deposition.

When partial pressure control of the reactive gas is used in conjunction with pulsed DC power, the shape of the hysteresis curve is quite different from when flow control is used. As is shown in figure A5.1.9 when the curve is plotted in the same manner as it is in figure A5.1.8, there is a negative slope region for the partial pressure between points A and B. In this region, the target is slowly but controllably poisoned. At point A, the deposition rate is at the full metal rate, whereas at point B the target is fully poisoned, and the rate is a small fraction of the metal rate. With the partial pressure control, all points are accessible between points A and B, and there are no forbidden compositions. Such control is possible with the combination of partial pressure control of the reactive gas and pulsed DC power (or medium frequency AC), which together prevent arcing on the target surface and thus maintenance of uninterrupted sputtering. Many of the other materials such as aluminum, zirconium, hafnium, or yttrium when reactively sputtered in an oxygen/argon atmosphere have a similar hysteresis curve as is shown above for the TiO_x system.

Recently Schneider *et al* reported a high rate deposition process for alumina [4]. Fig A5.1.10 shows the part of the hysteresis curve at the nose of the hysteresis curve which was studied in the work by Schneider *et al* . Oxide films produced at 0, 0.36, 0.41 and 0.43 mTorr O_2 partial pressure resulted in 100%, 92%, 76% and 38% deposition rates relative to the metal deposition rate. The accuracy of the reactive gas partial pressure control is better than 0.005 mTorr. The deposition rate in a pure Ar discharge is 9.9 Å s^{-1}.

The rate data for the AlO_x is given in figure A5.1.10. As expected, increasing the O_2 partial pressure reduces the deposition rate. The films produced at 76% and 38% deposition rate relative to the metal deposition rate appeared transparent, whereas the films produced at 92% were brown. These experiments showed clearly that this high rate can only be achieved by precisely maintaining a certain O_2 partial pressure value (position on the hysteresis curve). A shift towards higher O_2 partial pressures drastically reduces the deposition rate. An O_2 partial pressure increase of 5% (figure A5.1.10) decreased the deposition rate from 76% to 38% of the metal deposition rate. Conversely, a shift towards lower O_2 partial pressures resulted

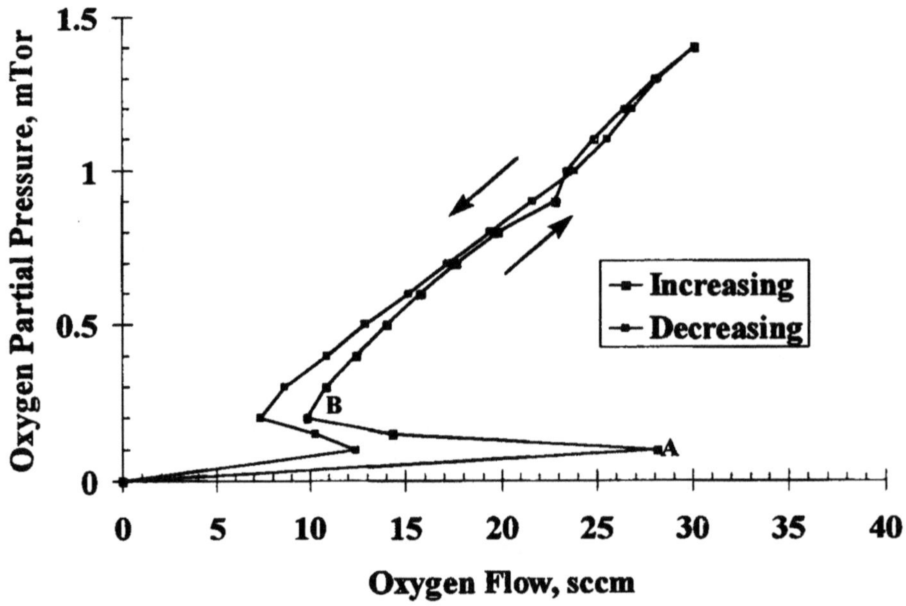

Figure A5.1.9. Hysteresis curve for the reactive pulsed DC magnetron sputtering of titanium in an argon/oxygen atmosphere using partial pressure control of the reactive gas [59].

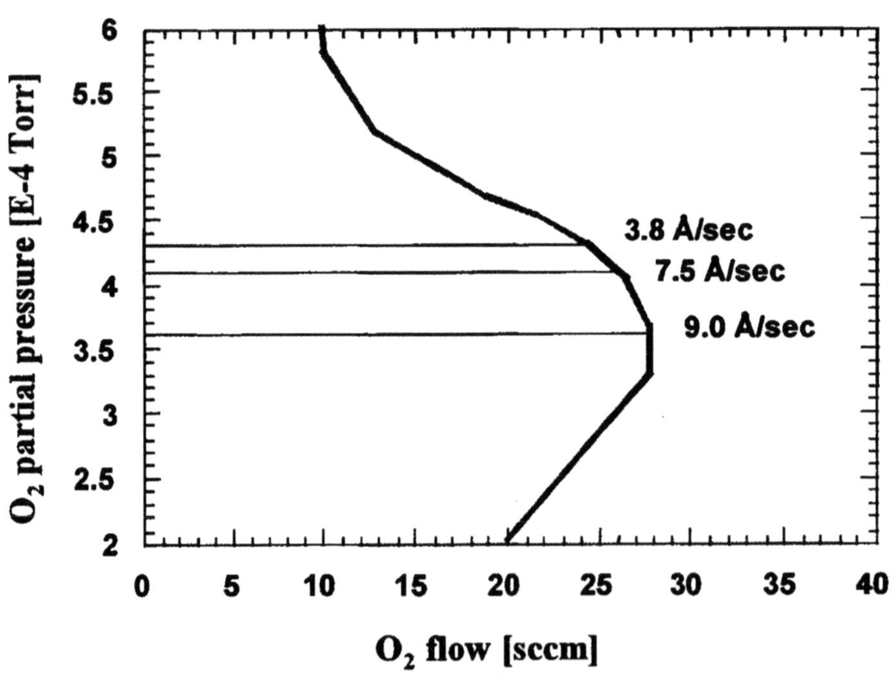

Figure A5.1.10. AlO_x hysteresis curve at 3 kW target power, 1 kW RF substrate bias power and 5 mTorr total pressure [4].

in the incorporation of free Al in the film. A reduction of 12% in O_2 partial pressure from the 76% rate value resulted in a film of low transmission and of lower hardness. To the best of the authors' knowledge at this time, 76% of the metal rate is the highest rate ever obtained by reactive pulsed DC magnetron sputtering of transparent AlO_x. In figure A5.1.11, the relative deposition rates with respect to the metal deposition rate (which is the highest that can be achieved for a given input power) of RF sputtering [17], DC sputtering with baffles [17], pulsed DC sputtering [52], and pulsed DC sputtering with partial pressure control are compared [4]. It can be seen that optimum high rate results can be achieved by utilizing both, medium frequency pulsed power and partial pressure control of the reactive gas. This technology was also utilized for the high rate deposition of close to stoichiometric zirconia films at a deposition rate relative to the metal rate of 82% [53].

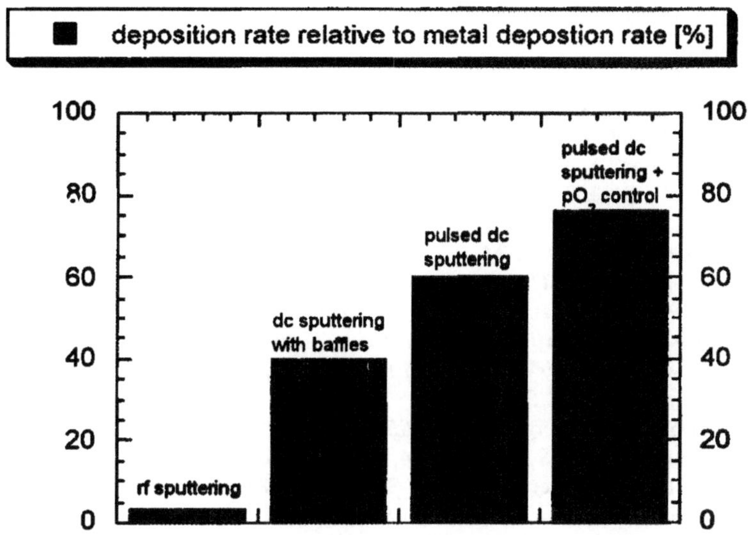

Figure A5.1.11. Relative deposition rates for RF sputtering [17], DC sputtering with baffles [17], pulsed DC sputtering [52], and pulsed DC sputtering with partial pressure control[4] are compared [57].

Recent modeling results by Macak [54] suggest that active process control is needed for the high rate deposition of alumina by magnetron sputtering if large target power settings and short source to substrate distances are employed. These modeling results are consistent with what has been observed experimentally when pulsed DC power and closed loop control of the reactive gas partial pressure is used.

A5.1.4 SUMMARY

Over the years there have been many schemes used to try to control the reactive sputtering process. Some of these have been successful for conducting films such as TiN, but for non-conducting films such as aluminum oxide the deposition process was much more difficult. It was not until two important advances in deposition technology—namely pulsed DC power and high speed process control techniques, for example, closed-loop partial pressure control of the reactive gas—were used together that it became practical to deposit non-conducting films at high deposition rates with controlled composition. Pulsed DC power and reactive gas partial pressure control are a very powerful combination that will expand the deposition and use of non-conducting films significantly.

REFERENCES

[1] Holleck H 1986 *J. Vac. Sci. Technol.* A **4** 2661–9

[2] Helmersson U, Todorova S, Barnett S A, Sundgren J-E, Markert L C and Greene J E 1987 *J. Appl. Phys.* **62** 481–4

[3] Sproul W D 1986 *J. Vac. Sci. Technol.* **4** 2874–8

[4] Schneider J M, Sproul W D, Chia R W, Wong M-S and Matthews A 1997 *Surf. Coatings Technol.* **96**

[5] Sproul W D 1996 *Science* **273** 889–92
[6] Thornton J A 1982 *Deposition Technologies for Films and Coatings: Development and Applications* (Park Ridge, NJ: Noyes)
[7] Hollands E and Campbell D S 1968 *J. Mat. Sci.* **3** 544–52
[8] Stirling A J and Westwood W D 1970 *J. Appl. Phys.* **41** 742
[9] Heller J 1973 *Thin Solid Films* **17** 163–76
[10] Gornachev B, Orlinov V and Popova V 1976 *Thin Solid Films* **33** 173–83
[11] Gornachev B, Orlinov V, Tsaneva V and Petrov I 1978 *Thin Solid Films* **52** 365–71
[12] Geraghty K G and Donaghey L F 1976 *J. Electrochem. Soc.: Solid State Sci. Technol.* **123** 1201–7
[13] Schiller S, Heisig U, Goedicke K, Schade K, Teschner G and Henneberger J 1979 *Thin Solid Films* **64** 455–67
[14] Sproul W D 1983 *Thin Solid Films* **107** 141–7
[15] Sproul W D and Thomashek J A 1984 US Patent No: 4,428,411
[16] Boxman R L, Martin P J and Sanders D M 1995 *Handbook Vac. Arc Sci. Technol.* (Park Ridge, NJ: Noyes)
[17] Jones F and Logan J 1989 *J. Vac. Sci. Technol.* A **7** 1240–7
[18] Kumagai H Y 1972 *IEEE Trans. Parts, Hybrids and Packaging* **PHP-8** 7–10
[19] Cormia R L, Tsujimoto N and Anderson S 1977 US Patent No: 4,046,659
[20] Kirchhoff V 1986 DD-Patent No: 252205 A1
[21] Muenz W D, Peterson H and Scherer M 1988 DE-Patent No: 38 02 852 C2
[22] Este G and Westwood W D 1988 *J. Vac. Sci. Technol.* A **6** 1845–8
[23] Scherer M, Schmitt J, Latz R and Schanz M 1991 *J. Vac. Sci. Technol.* A **10** 1772–6
[24] Scholl R 1997 *Surf. Coatings Technol.* **93** 7
[25] Ridge M I and Howson R P 1982 *Thin Solid Films* **96** 121–7
[26] Sproul W D, Rudnik P J and Graham M E 1989 *Surf. Coatings Technol.* **39/40** 355–63
[27] Valetta R M, Perri J A and Risman J 1966 *Electrochem. Technol.* **4** 402–6
[28] Stirling A J and Westwood W D 1971 *Thin Solid Films* **7** 1–10
[29] Greiner J H 1971 *J. Appl. Phys.* **42** 5151–5
[30] Thornton J A 1979 *US Dept of Energy Report DSE-4042-T-19*
[31] Chapin J and Condon C R 1979 US Patent No: 4,116,784, 1–10
[32] Schiller S, Beister G, Buedke E, Becker H-J and Schicht H 1982 *Thin Solid Films* **96** 113–20
[33] Enjouji K, Murata K and Nishikawa S 1983 *Thin Solid Films* **108** 1–7
[34] Penfold A S 1986 *Proc. Am. Soc. of Vacuum Coaters Annual Technical Conf., New Orleans* p 381
[35] Kadlec S, Musil J and Vyskocil J 1987 *Vacuum* **37** 729
[36] Berg S, Blom H-O, Larsson T and Nender C 1987 *J. Vac. Sci. Technol.* A **5** 202
[37] Berg S, Larsson T, Nender C and Blom H-O 1988 *J. Appl. Phys.* **63** 887
[38] Schiller S, Heisig U, Steinfelder K and Struempfel J 1979 *Proc. Int. Conf. on Ion Plating and Allied Techniques, London* (Edinburgh: CEP Consultants)
[39] Maniv S, Miner C and Westwood W D 1981 *J. Vac. Sci. Technol.* **18** 195–8
[40] Este G and Westwood W D 1984 *J. Vac. Sci. Technol.* **2** 1238–47
[41] Scherer M and Wirtz P 1984 *Thin Solid Films* **119** 203–9
[42] Pang T M, Scherer M, Heinz B, Williams C and Chaput G N 1989 *J. Vac. Sci. Technol.* A **7** 1254–8
[43] Aronson A J, Chen D and Class W H 1980 *Thin Solid Films* **72** 535
[44] Sproul W D 1984 US Patent No: 4,428,812
[45] Billard A and Frantz C 1996 presented at *ICMCTF, San Diego*
[46] Schiller S, Heisig U, Steinfelder K, Strumpfel J, Friedrich A and Fricke R 1987 *Proc. Int. Conf. on Ion Plating and Allied Techniques* (Edinburgh: CEP Consultants)
[47] Howson R P, Spencer A G, Oka K and Lewin R W 1989 *J. Vac. Sci. Technol.* A **7** 1230–4
[48] Affinito J and Parson R R 1983 *J. Vac. Sci. Technol.* A **2** 1275–1284
[49] Sproul W D 1987 *Surf. Coatings Technol.* **33** 73–81
[50] Sproul W D, Rudnik P J and Gogel C A 1989 *Thin Solid Films* **171** 171
[51] Rossnagel S M 1988 *J. Vac. Sci. Technol.* A **6** 19–24
[52] Schiller S, Goedicke K, Reschke J, Kirchhoff V, Schneider S and Milde F 1993 *Surf. Coatings Technol.* **61** 331–7
[53] Schneider J M, Sproul W D, Lefkow A R T, Matthews A, Graham M E and Rechner J 1996 *39th Ann. Tech. Conf. Proc., May 5–10, Philadelphia, PA* pp 168–73
[54] Macak K 1998 *MSc Thesis* Linkoping University
[55] Sellers J 1996 *Asymetric Bipolar Pulsed DC* ENI Technical Note
[56] Sproul W D, Graham M E, Wong M S, Lopez S, Li D and Scholl R A 1995 *J. Vac. Sci. Technol.* **13** 1188–91
[57] Schneider J M 1997 *PhD Thesis* University of Hull
[58] Sproul W D, Rudnik P J, Graham M E, Gogel C A and Mueller R M 1989 *Surf. Coatings Technol.* **39/40** 499
[59] Sproul W D 1996 *Materials Res. Soc. Symp. Proc.: Layered Materials for Structural Applications* ed J J Lewandowski, C H Ward, M R Jackson and W H Hunt Jr (Pittsburgh, PA: Materials Research Society) **434** 47

A5.2 AC and RF Reactive Sputtering

Grant O Este and William D Westwood

A5.2.0 INTRODUCTION

Sputtering is a low pressure physical vapor deposition process where ions are accelerated from a plasma across a potential drop to bombard the sputtering target. These energetic particles cause the ejection of neutral particles from the target which travel to the substrate located near the target where they condense to form a film.

Alternating Current (AC) sputtering is a technique where an alternating current is applied between two sputtering targets independently from ground. At frequencies of a few tens of kilohertz this process is essentially DC sputtering with the anode and cathode exchanging roles on each half cycle. Thus, each target is sputtered for almost 50% of the AC cycle. AC sputtering is characterized by a complete absence of DC current flow through the target and no current flow through the chamber walls to ground. This makes it possible to sputter conducting targets which are covered with a thin insulating layer but not a target which is itself an insulator (table A5.2.1).

Table A5.2.1. Comparison of current flows and target characteristics for reactive sputtering systems.

	AC	RF	DC	Pulsed DC
Net DC current flow	No	No	Yes	Yes
Current to ground	No	Yes	Yes	Yes
Anode required	No	No	Yes	Yes
Targets	dual	single/dual	single	single/dual
target type	conducting	insulating	conducting	conducting
poisoned target mode	stable	stable	arcing	arcing?

In Radio Frequency (RF) sputtering, the frequencies are normally greater than 1 MHz. In the usual RF configuration, the RF power is capacitively coupled through a matching network to a single target with the return current path to ground through the ground electrode and the chamber walls. Since the ions in the plasma are unable to follow the electric fields at these frequencies, but the electrons can, the dark space between the plasma and the target becomes a rectifier. This rectifying action produces a negative DC bias on the surface of the cathode which causes an ion current to flow continuously to the target, so that sputtering is continuous. This makes it possible to sputter an insulating target (table A5.2.1); a conducting target with a thin insulating compound surface layer is a subset of a fully insulating target and can also be sputtered by RF. The RF supply can also be coupled between two targets, as in the AC case; although this is used in special applications, it is not the usual configuration.

These systems can operate with both diode and magnetron target assemblies. RF diode sputtering has been a standard method for sputtering dielectrics since the 1960s. A 60 Hz version of the AC was used by Kumagai [1] to sputter Ta and Al diode targets. However, magnetron systems are most popular for reactive sputtering because of the higher rates and only magnetrons are considered here.

Early work on AC magnetron sputtering of dielectrics by Cormia [2] used a single cathode powered with respect to an anode at frequencies above 10 kHz. Since the anode was negative for half of each cycle, it was alsp sputtered. The current mid-frequency (20–100 kHz) dual cathode arrangement was introduced by Este and Westwood in 1987 [3] and subsequently refined by Scherer *et al* [4] in 1992 and Glocker [5] in 1993.

A5.2.1 AC REACTIVE SPUTTERING

AC sputtering principles

In AC reactive sputtering [3–5], targets are powered in pairs by a balun transformer (figure A5.2:1) such that on each half cycle one target is the anode while the other is the cathode; on the next half cycle the roles are reversed. The frequency of these reversals is typically 40 kHz, although frequencies in the 20–100 kHz range have been used. AC sputtering is characterized by the complete absence of a net DC current flow through the targets or any current flow through the chamber walls to ground. The two targets could equally well be enclosed in a glass or other dielectric chamber with no reference to ground. This approach has several advantages over RF. First, under similar conditions, the rate is approximately double that in RF and close to the DC value [3]. Second, the equipment is simpler with no requirement for reactive matching networks. The power supply is basically a standard DC switching supply with the rectifiers removed. It is designed to deliver power into a 50 Ω load. A transformer is usually required to match the output voltage to the target characteristic so that maximum power can be delivered by the supply.

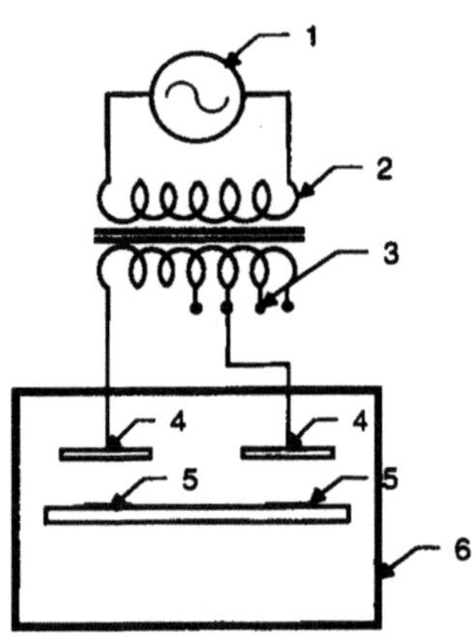

Figure A5.2.1. AC powering of dual targets. (1) AC power supply (2) Balun or matching transformer (3) transformer output taps for matching to (4) targets with isolated shields (5) substrate holder (6) vacuum chamber.

The AC process provides similar rates to DC sputtering but is not affected by thin insulating layers on the cathode surface because the two targets alternate between anode and cathode. The voltage waveforms on one of the targets of the pair shows the rectification [4] as well as the ignition of the plasma on each cycle [5]. The compound layer on each target, which is the 'anode' for half the time, is semi-continuously sputtered and therefore does not increase in thickness as it does on the anode in a DC system. For reactive sputtering this completely avoids the disappearing anode problem associated with DC. Even when the system is operated in the poisoned target mode the anode is effective because the current is capacitively coupled across the dielectric layer on both targets and there is no DC component to cause breakdown.

This capacitive coupling is frequency dependent, however, and at lower frequencies the current must be reduced to avoid dielectric breakdown. The frequency at which dielectric breakdown will occur can be calculated if the dielectric strength and dielectric constant (K) of the compound layer on the target is known. The surface of the target is then modeled as a simple parallel plate capacitor:

$$C = \varepsilon_0 K A / d \qquad \text{(A5.2.1)}$$

where ε_0 is the permittivity of free space, A is the area (in m^2) and d (in m) is the film thickness. For a given breakdown field (E_b) in the insulator, the breakdown voltage V_b is proportional to film thickness,

$$V_b = E_b \cdot d \tag{A5.2.2}$$

so that

$$C = \varepsilon_0 K A E_b / V_b \tag{A5.2.3}$$

The difficulty with this calculation is determining the appropriate values of K and E_b for the films on the surface of a sputtering target. These are probably inhomogeneous and sub-stoichiometric, because of the sputtering, and are quite thin (< 50 nm). Assuming the bulk values probably overestimates the value of C. Assuming 10^8 V m^{-1} for E_b and a dielectric constant of 3 for SiO_2, $C/A = 2.66 \times 10^{-11}$ farad $\cdot$ m^{-2}. Note that the thermal SiO_2 in silicon transistors is ~ 5 nm thick and sustains > 5 V; i.e. 10^9 V m^{-1}.

For a frequency f (Hz) the current (I) at breakdown is

$$I = 2\pi f C V_b \tag{A5.2.4}$$

which gives a current density of 667 A m^2 for 40 kHz. For higher frequencies, a higher current density can be reached before breakdown.

For a 15 cm diameter target this corresponds to a current of 16.5 A or a power of 8 kW for a typical magnetron target voltage of 500 V; however, the current density is not uniform over a magnetron target and is at least 10 times higher in the center of the erosion track, indicating that breakdown could begin at power levels of 0.8 kW. In the case of an unpoisoned target, most of the current flows in the metallic areas and the current density will be a factor of 10 lower on poisoned areas of the target, which will be at the edges of the erosion track. Cathodes with moving magnets or larger erosion areas will be better. Materials with higher values of K and E_b will have a higher current (i.e. sputter power density) rating before breakdown.

Sputtering elemental targets with a compound surface layer allows higher power densities since they are generally better thermal conductors than insulating targets and target cooling is improved. If the reaction occurs on the target, this thin compound layer is sputtered in the same way as a compound target. This mode of operation is commonly referred to as the 'poisoned target mode'. However, an insulating compound target (e.g. 1cm thick SiO_2) cannot be sputtered using these AC frequencies; the low capacitance in equation (A5.2.3) gives a very low current.

If the flow of the reactive gas is precisely controlled, it is possible in some cases to sputter from an unreacted target surface while supplying enough reactive gas to form a compound at the substrate. This is referred to as the 'unpoisoned' mode. Baffles and directed gas supplies are often used to separate the reactive gas in the substrate region from the target to prevent poisoning the target.

Scherer *et al* [4] used a feedback loop to adjust the reactive gas flow to maintain the target voltage at the value required to deposit nitride and oxide films on substrates moving past the targets. For Si targets sputtered in Ar/N_2, the target voltage (V_T) decreased monotonically from approximately 550 V to 500 V as the N_2 flow was increased and a nitride layer formed on the targets. For Al targets, V_T decreased discontinuously from > 600 V to < 400 V as the O_2 flow was increased and an Al_2O_3 layer formed on the target. The use of feedback control loops is discussed in section A5.1

If the compound forms on one target before the other, the target voltages will differ and the sputtering power will therefore also differ since the current is the same to both targets. Compound formation will occur more rapidly on the lower power target, making control more difficult. The power supply must also support the voltage difference.

Control is improved by sputtering at high power with high pumping speeds. At low power (< 1 kW) it becomes very difficult to control these reactions and background contamination becomes more important. At low power, oxide films have even been deposited from metallic targets without the addition of O_2, due to reaction with the background water vapour. High pumping speeds flatten out the hysteresis curve of reactive gas pressure versus gas flow [6]. This increases the control of the operating point on the hysteresis curve, as discussed in sections A5.1 and A5.3

One problem associated with AC sputtering is that it requires two or more targets. This is not a significant problem in large production coaters where several cathodes are operated side by side to provide the required throughput, but it is less convenient for smaller systems. The development of paired cathodes specifically designed for AC will reduce this problem; for example, two narrow rectangular cathodes mounted on a single plate would have dimensions similar to a larger single target and provide the same total sputtering area. Although not specifically designed for AC sputtering, circular concentric split cathode assemblies, originally designed to improve deposition uniformity, can be used in an AC mode.

Glocker [5] estimated from Langmuir probe measurements that there was additional substrate heating in AC sputtering due to electrons and ions from the plasma; the plasma density was a factor of 4 greater than for DC and the electron temperature was higher. However, compared with other reports, they measured quite low values of substrate heating for AlN deposition; the energy per deposited atom was 32 eV and 20 eV for AC and DC sputtering respectively compared with over 1000 eV per atom reported for AlN [7]. The latter high values were obtained for RF sputtering a nitride layer whereas the low values were obtained for sputtering the metal target and forming the nitride at the substrate. Szczyrbowski *et al* [8] suggested that dense oxide films resulted from high-energy ion bombardment of the substrate in the AC plasmas.

The pulsed DC technique discussed in section A5.1.can be considered as the superposition of AC and DC currents so that the target periodically goes positive and thus reduces arcing. In the bipolar pulsed version the targets are employed in pairs similar to the AC approach except that they are driven independently out of phase by separate supplies referenced to the chamber ground. The claimed advantage is that rates for the two targets can be adjusted independently. However, care must be exercised to avoid sputtering the chamber wall.

AC system requirements

A practical set-up for AC sputtering requires the following.

(1) A suitable vacuum chamber including pumping and vacuum gauging. Because there are no currents flowing through the chamber walls a glass bell jar system would be feaible but stainless steel systems are generally used. Most types of throughput pumps or cryopumps will be satisfactory. Throttling may be required although high pumping speeds improve the stability of the reactive gas control loop. Care should be exercised when pumping oxygen with pumps containing hydrocarbon oils, and also when using cryopumps due to the buildup of ozone in the pump. Guidance from the cryopump manufacturer on regeneration procedures should be obtained.

(2) Two magnetron targets installed in the chamber. The targets must be able to 'see' one another in a plasma sense. This means there must be a clear path between the two cathodes; approximate dimensions are 500 cm^2 in cross section with the minimum dimension being 10 cm but this will depend on the system and operating pressures. The targets should be the same size if they are to be operated under critical gas flow conditions, i.e. near a hysteresis transition point. Some nonuniformity can occur in the target erosion pattern due to the interaction of the plasma between the two cathodes but this affect is usually less than 10% across the target; it can be reduced by target rotation.

It is essential that there be no conducting surfaces, such as darkspace shields, penetrating the magnetic fields of the cathodes. Magnetron assemblies usually include ground space shields which often also serve as the anode for DC sputtering. In AC sputtering, these ground shields will intercept electron flow. If shielding is required, these 'ground' shields must be isolated from ground. Glass or alumina shields could be used or metal shields should be mounted on insulators. Even if conducting films are deposited on these shields, they should be arranged to remain at floating potential, separated from ground.

It is important to ensure that the power coupling to the targets is truly differential with no ground connections, including capacitive coupling to ground. A target must not inadvertently be shorted to ground. Although the system will appear to be working normally with one target grounded, the grounded surfaces, usually including the chamber walls, are receiving ion current and are thus being sputtered. For correct operation, the same AC voltage should appear between both targets and the chamber. Conical magnetrons ('S' gun type) with a central anode should have the anode tied to the cathode or left floating with respect to ground.

It is possible to run more than one pair of targets in the chamber at the same time. The two pairs may be run at the same frequency if they are synchronized; since there is no common current path there is little interference. However, this would be minimized if they are run at different frequencies, such as 35 kHz and 40 kHz; the 5 kHz beating between the two pairs being sufficiently rapid that it has little affect. If the supplies are phase-locked properly, it should be possible to keep a plasma ignited in the chamber continuously. However, there may be problems associated with controlling the reactive gas flows if the pairs of targets are different sizes or materials, or if different power levels are used for the two pairs.

(3) A balanced drive for the two targets is required. The simplest version consists of a balun transformer. This is a transformer which is fully isolated in terms of the primary to the secondary windings, and the secondary winding is simply floating with respect to ground. This transformer can also be used to accomplish the impedance matching between the plasma and the power supply by employing a 50 Ω primary and a multi-tapped secondary winding (figure A5.2.1) One target is then driven from one end of the secondary and the other from the appropriate secondary tap. This transformer should be located as close to the targets as possible. If shielded cables are used care should be taken that the shields are not grounded at both ends as substantial eddy currents can be induced in these ground loops. It is recommended that the conductors between the transformer and the targets be kept as close together as possible, by twisting where applicable. Commercial power supplies often have differential outputs fully isolated from ground so that the balun transformer is unnecessary.

(4) A suitable power supply. With the use of the balun transformer it is possible to use any single-ended power supply; a variable frequency may be useful in a research environment. However, auxiliary equipment will be required for the measurement of the sputtering current or cathode voltage. Fixed frequency supplies with fully floating differential outputs and complete power, current and voltage monitoring are generally more convenient. A load match transformer is required to allow efficient coupling of power to the targets under different conditions.

When an elemental target is poisoned, there is a rapid change in target voltage [4] and the matching is required to maintain a constant power. The change in V_T for Nb as O_2 is added to the system is shown in figure A5.2.2 for both DC and AC operation at approximately 3.5 A; the current from the DC supply is fixed but it varies slightly ($< 5\%$) in the AC case due to the matching. For the unpoisoned target, V_T is higher for the AC case at low flow rates but the DC and AC values are similar near the flow for target oxidation. When the target surface oxidizes, V_T decreases and the current increases; since the sputtering rate is lower for the oxide, the oxygen flow has to be reduced to return to the unpoisoned mode. For Si targets, V_T decreases from about 1000 V with increasing oxygen flow; just prior to target poisoning, the value may vary between 950 and 650 V depending on the matching conditions and decreases to about 400 V on poisoning.

(5) A suitable substrate support to hold the substrate or substrates. The wide range of applications for sputter coating involves an equally wide range of substrates, sizes and types and it is impossible to cover all the different types of fixturing used to support the substrates in the correct position with respect to the targets. The simplest fixture might be a clip holding a piece of glass whereas the more complex one might involve moving cathodes on complex articulated arms to follow the profile of a molded aircraft canopy. In some cases, substrates may have to be heated or cooled to obtain the required film properties.

(6) A thickness monitor is useful for determining rates during the deposition cycle. This unit should be made specifically for sputtering applications. For non-reactive and poisoned target modes the product of power and time for the process will usually yield thickness control to $\pm 3.0\%$. The conversion of an elemental target to the poisoned mode is usually evident from the decrease in deposition rate per kW.

(7) Gas flow controllers for the reactive gas, and the non-reactive gas when mixtures are used. The gas pressure required for running AC is typically higher than for DC, and the plasma ignition also requires a higher pressure, or another source of ignition. For the same magnetron assembly and target in Ar which operates stably in DC at 0.5 mTorr the pressure for stable AC is about 1 mTorr. Whereas the DC plasma will ignite at 0.5 mTorr, the AC may require a pressure of several mTorr; the pressure can then be reduced

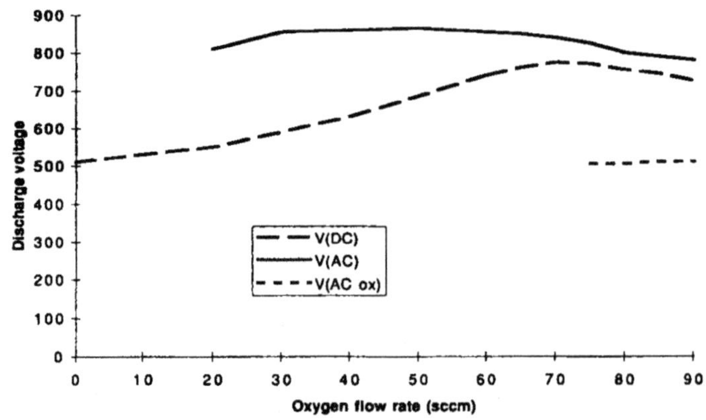

Figure A5.2.2. Discharge voltage at a current of 3.5 A for AC and DC powered Nb targets as a function of O_2 flow; the value for an oxidized target with AC power is also given.

where required for the deposition process. AC supplies typically do not provide as high a voltage for ignition.

Although a target acts as cathode and anode for only 12.5 μs during every 40 kHz cycle, the plasma does not completely extinguish in this time. Significant numbers of ions and electrons are available to initiate the next half cycle of the discharge. Similarly, the plasma can be ignited at the operating pressure by an alternative source of electrons, such as a hollow cathode. A magnetron which is already operating is also a suitable ignition source.

Deposition rates

The sputtering rates depend on the power coupled to the magnetrons and the state of the target. If the target is 'poisoned' the rates are typically significantly lower, as shown in table A5.0.1. To obtain maximum deposition rate, the target should be sputtered fast enough to prevent compound formation on the target. However, good control of the reactive gas flow is necessary to maintain this condition. For process stability, sputtering from a poisoned target is preferable; lower reactive gas flows and pumping speeds can be used since the reactive gas has to be supplied only to replace that sputtered from the target. However, the deposition rates are lower and this can be a serious affect; for example, the rate for Al_2O_3 is approximately 15–20 times slower from a poisoned target.

If a SiO_2 target is sputtered at a power of 5 kW and the erosion track area is 50 cm^2 (e.g. a circular target with 3 and 5 cm radii of erosion track), the sputtering rate is 0.05 nm s^{-1}, assuming uniform sputtering over the erosion track, and 1×10^{19} O atoms $\cdot$ s^{-1} must be replaced on the target surface (i.e. 2×10^{17} O atoms $\cdot$ cm^{-2}s^{-1}). An O_2 flow rate of 1 sccm provides $4.5 \times 10^{17} O_2$ s^{-1} to the system so that a flow of at least 11 sccm is required; the actual flow required will be higher since some O_2 will be pumped away before reacting.

For Al in the unpoisoned mode, an O_2 flow of 85 sccm is calculated from table A5.0:1 for a sputtering power of 4 kW whereas for a poisoned target the required flow is 4.2 sccm. Experimental values for the latter case gave 15 sccm of O_2; about 25% of the O_2 would be used and the remainder pumped away.

A5.2.2 RF REACTIVE SPUTTERING

RF Principle

Dielectrics have usually been sputtered from a single dielectric target using RF, usually at 13.56 MHz. A schematic diagram of an RF sputtering system is shown in fig A5.2.3 The RF power supply is connected through a matching network which transforms the reactive plasma impedance into a 50 Ω resistive load. The matching network powers the cathode with respect to ground potential, which in general is the vacuum chamber. The sputtering gas is typically argon with the addition of the reactive gas to compensate for losses occurring during the sputtering process. For example, if an oxide target is sputtered, O_2 is added to obtain stoichiometric films. If the sputter rate is low and the process system requires opening of the system to load and unload substrates, the residual background water vapour may provide sufficient oxygen for this purpose.

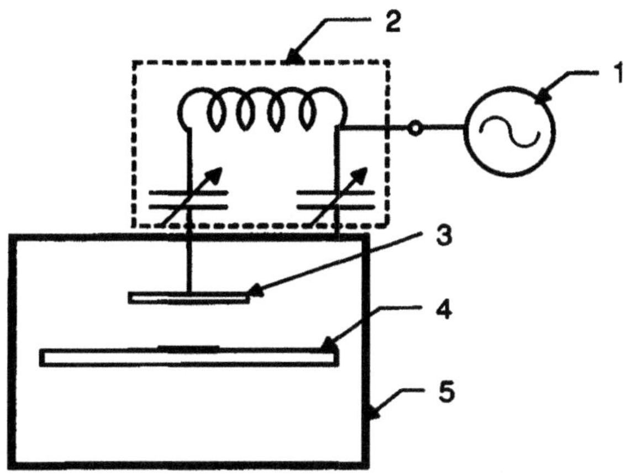

Figure A5.2.3. RF powering of an insulating target: (1) RF generator (2) automatically tuning matching network (3) target (4) substrate holder (5) vacuum chamber providing ground return to generator.

The target for RF sputtering is usually a solid insulator; e.g. a 1 cm thick SiO_2 slab. The RF supply provides capacitive coupling through a thick insulator. Using equations (A5.2.1)–(A5.2.3), the current for breakdown is about 300 times that for AC sputtering; i.e. RF sputtering is not limited in power by dielectric breakdown but by the limitations of RF supplies.

Dielectric films can also be deposited by RF sputtering an elemental target (e.g. Al) in a reactive gas (e.g. O_2) so that the compound (e.g. Al_2O_3) is formed on the substrate. Elemental targets are generally better thermal conductors so that target cooling is improved.

RF sputtering has two main drawbacks compared to AC. First, the rate is approximately 50% of the DC rate under equivalent conditions [3] and this loss in efficiency is due in part to power being dissipated in plasma heating; this heats the substrates in addition to the heating due to the sputtered atom energy. Second, RF power generators and the requirement for complex matching networks makes the cost per watt of sputtering power double that of a DC or AC generator. The RF generator is designed to deliver power into a 50 Ω load; the matching network has to be tuned to provide the impedance match and thus minimize the power reflected to the RF supply.

The big advantage of RF sputtering is its universality. Any target material can be sputtered using RF essentially without arcing and this produces lower particulate counts in the films. However, this advantage can easily be thwarted by particles being released from fixturing due to differential thermal expansion or mechanical vibration.

RF System requirements

A practical set-up for RF sputtering requires the following. Many of the considerations are similar to those for AC sputtering.

(1) A suitable vacuum chamber including pumping and vacuum gauging. A glass bell jar system is sufficient for RF but care must be exercised to insure that there is a sufficiently large effective area associated with the grounded surfaces relative to the target. The 'effective' area of the grounded surfaces

(i.e. base plate, grounded fixturing, etc) is not well defined since the plasma will not penetrate into all the nooks and crannies of complex fixturing. This problem is exacerbated by the fact that the degree of penetration will depend on the pressure in the system and the sputtering atmosphere. The ratio of ion energies at the target and ground surfaces is proportional to the ratio (ground area/target area)^n, where $n \sim$ 2–4. As the area ratio approaches 1, the grounded surfaces will be sputtered by ions with the same energy as those arriving at the target. An area ratio of at least 4 should reduce sputtering of the ground surfaces to negligible values.

Of course, two identical targets can be coupled to the RF supply just as in the AC configuration. This 'facing target' arrangement has been used to sputter targets from which O^- ions are sputtered (e.g. ZnO, $YBa_2Cu_3O_6$) to avoid O bombardment of substrates.

A metal vacuum chamber is to be preferred. Most types of throughput pumps are satisfactory. Throttling may be required although high pumping speeds improve the stability of the gas control loop. Care should be exercised when pumping oxygen with pumps containing hydrocarbon oils or with cryopumps.

(2) An RF magnetron cathode must be installed in the chamber. In general there is little difference between RF and DC cathodes. The primary differences are due to the inductance of the current path and the degree of shielding and the spurious capacitance of the shielding. More than one cathode can be installed and operated in the chamber at the same time but care must be exercised to phase lock the power supplies to prevent beating since they are driven with respect to a common ground.

(3) An RF matching network will be required to convert the plasma impedance as seen through the cathode to the 50 Ω required by the RF power supply. The automatic matching networks are to be preferred especially when processes are running near the transition from elemental to poisoned mode; the target voltage changes rapidly and the matching network must adjust to maintain constant power. The unit should be installed so that the connections to the chamber and the cathode are as short as possible. Substantial circulating currents will be present during operation and so the conductors used should be substantial. This problem is exacerbated because of skin effects and so materials such as copper tubing or braid should be employed.

Since insulating surfaces can couple RF power out of the plasma, creating high field strengths in the vicinity of windows, it is possible to inadvertently radiate substantial amounts of RF energy and it is prudent to perform a RF field strength survey to ensure that levels are below North American guidelines for radio frequency exposure [9, 10].

(4) There are many manufacturers of suitable power supplies for sputtering. They are usually solid state units operating at 13.56 MHz and, unlike the older tube types, are very sensitive to reflected power. Without the automatic matching networks it may be quite difficult to ignite the plasma. Once again one should use as much power as practical. Since only 50% of the power is dissipated in the target [3] the power supply rating should be higher than the required sputtering power.

(5) Suitable substrate support to hold the substrate(s). The considerations of size, etc, are the same as for AC systems.

(6) A thickness monitor is useful for determining rates and thicknesses during the deposition cycle. This unit should be specifically for sputtering applications. Problems with noise are often encountered when using these types of sensors in conjunction with high RF fields. For non-reactive and poisoned target modes a simple power times time for the process will usually yield thickness control to ± 3.0%.

(7) Gas flow controllers for the reactive gas, and the non-reactive gas when mixtures are used.

A5.2.3 FUTURE DIRECTIONS

AC sputtering systems have gradually been introduced over the last decade and new applications and methods are likely to be developed with experience. One area which has not yet been exploited is that of polyphase drives. For example, it should be possible to drive three cathodes using three power supplies running at the same frequency with fixed 120 degree phase differences between them so that the power to each target can be varied independently over a limited range, making it possible to adjust the deposition uniformity continuously. It may be useful to change the relative phasing of the cathodes to vary the relative rates. This can be achieved without any current flow through the chamber and no net DC current flow through any of the targets. However, this requires special types of power supply.

Using the above polyphase type of set up it would also be possible to apply an AC bias to the substrates enhancing ion bombardment and thus modifying the growing film properties. Unlike a DC bias the AC will couple through a dielectric without the risk of damage to the film or substrate.

New AC power supplies are becoming available which are intended to allow two targets of different materials to be sputtered. However, targets of different materials have different operating conditions (e.g. voltage). If there are no additional current paths in such a system, then the same current must pass through both targets. If the current on each half cycle is varied to achieve the required rates, a net DC current may flow through the cathodes increasing the risk of dielectric breakdown. The voltage applied to different targets is simply determined by the plasma impedance for each target. If the different currents to the targets are balanced by using the chamber as a third electrode, the vacuum chamber itself cmay be sputtered, possibly resulting in film contamination.

RF sputtering has been in use since the 1960s and magnetrons since the 1970s so that RF magnetron sputtering is a quite mature technology. However, new applications to reactive sputtering for specific applications can be expected.

REFERENCES

[1] Kumagai H 1972 IEEE PHP-8, p 7

[2] Cormia R 1977 US patent no. 4,046,659

[3] Este G O and Westwood W D 1988 *J. Vac. Sci. Technol.* A **6** 1845

[4] Scherer M, Schmitt J, Latz R and Schanz M 1992 *J. Vac. Sci. Technol.* A **10** 1772

[5] Glocker D A J 1993 *J. Vac. Sci. Technol.* A **11** 2989

[6] Okamoto A and Serikawa T 1986 *Thin Solid Films* **137** 143

[7] Este G O and Westwood W D 1987 *J. Vac. Sci. Technol.* A **5** 1892

[8] Szczyrbowski J, Brauer G, Ruske M, Teschner G and Zmelty A 1997 *J. Non-Cyrstalline Solids* **218** 262

[9] IEEE 1992 C95.1: Standard for Safety Levels with Respect to Human Exposure to Radio Frequency Electromagnetic Fields, 3 kHz to 300 GHz

[10] IEEE 1992 C95.3: Recommended Practice for the Measurement of Potentially Hazardous Electromagnetic Fields—RF and Microwave

A5.3 Modeling of the Reactive Sputtering Process

S Berg, T Nyberg, H-O Blom and C Nender

A5.3.0 INTRODUCTION

Manually operated coating processes are becoming quite rare in modern material fabrication. Small tolerances and demands to operate at optimal processing conditions call for very precise control of the various parameters. To be able to operate a process at optimal conditions, an in-depth understanding of the process is necessary. Only a full understanding of the behavior of the process will make it possible to judge if the selected set of parameters will result in the desired optimal processing condition.

Process modeling, often referred to as Technology-CAD, has become a powerful technique available to the processing industry for process control and optimization. Having access to a reliable theoretical model for a particular process will make it possible to carry out process simulation and ultimately to predict the outcome from a specific set of parameters. Experimentally it is normally not possible to independently vary all the parameters involved. In theoretical process modeling, however, this can be easily done. By carrying out systematic process modeling studies it is often possible to obtain a detailed understanding of the process behavior in quite a short time. On the other hand, to obtain the corresponding results by experimental studies may take weeks or months. Introducing process modeling may therefore save time (and money) in production as well as giving valuable help in finding appropriate processing conditions.

Sputtering is an example of a widely used industrial thin film coating process. By extracting positive ions from a glow discharge (normally inert argon gas) it is possible to sputter atoms from the negatively biased target (cathode) and form a thin film of this material on the surrounding surfaces. Basically this process is quite simple and straightforward. The inert gas does not react with the sputtered material. Furthermore the film composition will mostly be identical to the composition of the cathode. This 'insensitivity' to processing conditions has made sputter deposition a widely popular and used technique in the coating industry.

Sputtering is also a very flexible coating process. By adding e.g. a small amount of oxygen to the argon it is possible to induce reactions between the sputtered material and the oxygen molecules. In this way it is possible to form oxides by sputtering elemental metal atoms from the cathode in an argon/oxygen mixture. This process is called *reactive sputtering* [1]. Despite the similarities, the reactive sputtering process behaves dramatically different from the basic inert sputtering process. The primary cause of this big difference in behavior is that the reactive sputter deposition process is extremely sensitive to the partial pressure of the reactive gas. This partial pressure is influenced by both the electric power supplied to the cathode and of course to variations in the supply of the reactive gas. At first sight this may not seem to be a big problem. However, the way these parameters interact causes the reactive sputter deposition process to behave very non-linearly and in most cases also to exhibit hysteresis effects.

It should be noted that in reactive sputtering it is not enough to know the values of the desired processing parameters. The sequence in which these parameters are set may also be of vital importance. That is, one may not necessarily reach the same processing conditions, by first adjusting the supply of the reactive gas and then the electric power fed to the target as opposed to the reversed sequence. It may well happen that, in the first case, one deposits a compound material while, in the second case, the metal will be insufficiently reacted and a metal rich coating will be deposited. This effect makes it a somewhat delicate matter to apply process control to reactive sputtering processes.

We hope that the description of the reactive sputtering process given below sufficiently clarifies its behavior and serves to guide the reader to successfully design several reactive sputter deposition processes.

A5.3.1 MODELING OF THE REACTIVE SPUTTERING PROCESS

Here we will outline a simple, but for most applications quite satisfactory, model for the basic reactive sputtering process referred to as 'Berg's model', first suggested 1987 by Berg and his co-workers at Uppsala University [2]. It has been proven to fit most experimental findings concerning the reactive sputtering process found in the literature [3]. From this model it is possible to predict the drastic variations in the deposition rate, the composition of the deposited film, the partial pressure of the reactive gas, the 'poisoning' of the target surface by the reactive gas and the position as well as the width of the hysteresis region. The model can easily be modified to include several targets (reactive co-sputtering) or reactive sputtering from an alloy (composite) target [4]. It can also be extended to include more than one reactive gas (e.g. oxy-nitrides) [5].

To be able to obtain a simple and instructive model, however, it is necessary to make a few simplyfying assumptions. It is up to the reader to improve the model at a later stage by modifying some of these basic assumptions and develop a somewhat more sophisticated model. However, referring to figure A5.3.1 below, we use the following assumptions.

- The target is an elemental metal material (Al, Cu, Ti etc).
- Only ions from the inert gas contribute to sputter erosion of the target. The sputter contribution by ions from the reactive gas is neglected.
- Compound molecules sputtered from the target surface are deposited as molecules.
- The target current is totally carried by the ions from the inert gas. The contribution from emitted secondary electrons from the target is neglected.
- The ion current at the target is uniformly distributed over the target surface (A_t).
- All the sputter eroded material from the target surface is uniformly collected at the 'collecting' substrate surface A_s. This collecting surface represents both the substrate surface and all other receiving surfaces in the vacuum chamber.
- During processing a fraction of the target metal surface will react with the reactive gas and form the corresponding metal compound. The fraction of the target covered by this compound is denoted Θ_t. In figure A5.3.1 this is illustrated by dividing the target area A_t into two parts: the surface fraction Θ_t consisting of the metal-compound and the surface fraction $(1 - \Theta_t)$ consisting of the 'fresh' unreacted metal.
- The same reaction takes place at the collecting area A_s. However, the fraction Θ_s covered by the compound material (= reacted target metal) at this surface may differ from the fraction Θ_t at the target surface.
- The reactive gas supplied to the processing chamber gives raise to a partial pressure, P_r, that is evenly distributed in the chamber. This pressure causes a flux, Γ_r, of reactive gas molecules to all surfaces in the chamber. The relation between P_r and Γ_r is

$$\Gamma_r = \frac{P_r}{\sqrt{2\pi mkT}} \qquad \text{(A5.3.1)}$$

  where m is the mass of the reactive gas molecule, k is the Boltzmann constant and T the absolute temperature.
- The sticking coefficient is defined as the probability of a reactive gas molecule to stay on a surface when impinging on it. This probability is set to zero for the compound covered parts of the chamber and target surfaces.

Based on the assumptions above it is possible to define three basic balance equations describing this 'simplified' reactive sputter deposition process.

Conditions at the target area A_t

Compound formation at the target surface will only take place due to reactions between the reactive gas and the fraction $(1 - \Theta_t)$ of the target consisting of the unreacted target metal. The rate of compound

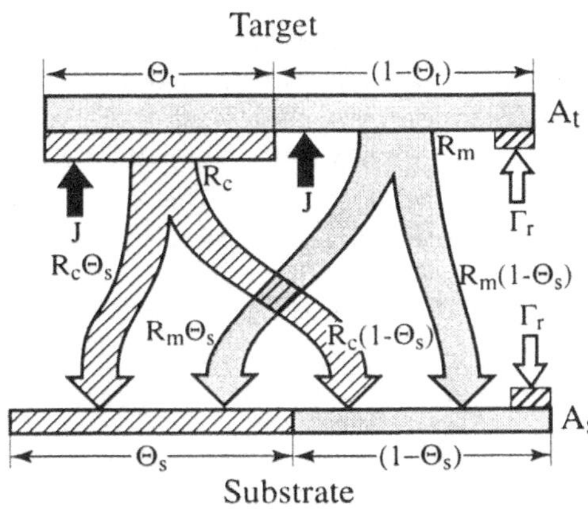

Figure A5.3.1. A schematic of the transport of the sputtered material from the target to the substrate, reactive gas fluxes and ion fluxes in a reactive sputtering process. The notations are defined in the text.

formation (number of molecules per unit time) at the target may be written as:

$$\Gamma_r \alpha_t (1 - \Theta_t) A_t b$$

where α_t is the sticking coefficient of the reactive gas molecules at the metal surface and b is a factor that defines the number of compound molecules formed by one reactive gas molecule.

The rate at which compound molecules are sputtered from the target surface can be written as

$$\left(\frac{J}{q}\right) \Theta_t A_t \gamma_c$$

where J is the ion current density (normally argon ions) at the target surface, q is the elementary charge, $(\Theta_t A_t)$ is the area of the target surface covered by the metal compound and γ_c is the sputtering yield of the compound for an incident energetic ion. Note that γ_c depends on the energy of the ion [6, 7].

At steady state the compound formation rate (due to reactions between the metal target and the reactive gas) must be identical to the sputter erosion rate of compound material from the target. This means that Θ_t has reached its steady state value. This gives us the first basic balance equation

$$\Gamma_r \alpha_t (1 - \Theta_t) A_t b = \left(\frac{J}{q}\right) \Theta_t A_t \gamma_c \tag{A5.3.2}$$

From equation (A5.3.2) in conjunction with equation (A5.3.1) it is possible to calculate Θ_t for any given value of P_r since all other parameters in the equation can be defined by the user.

The total sputter erosion rate R from the target can be expressed by

$$R = \left(\frac{J}{q}\right) \left[\gamma_c \Theta_t + \gamma_m (1 - \Theta_t)\right] A_t \tag{A5.3.3}$$

where γ_m is the sputtering yield of metal atoms due to sputtering by the energetic ions. The erosion rate here is expressed in units particles (metal atoms and compound molecules) per second. This could easily be converted into other units if desired.

Conditions at the collecting area A_s

A similar argument to that for the conditions at the target can be used for defining the conditions at the collecting area A_s. At steady state the rate of compound formation must be identical to the rate of 'compound elimination' from this surface. No sputter erosion of compound material, however, takes place from this surface. Instead 'compound elimination' is carried out by the elemental metal atoms sputtered from the target which are deposited onto the compound fraction Θ_s of the collecting surface A_s. Due to this some 'fresh unreacted metal' is added on top of the compound covered fraction Θ_s of the collecting surface thereby converting it to a 'fresh metal' surface. This effect causes a decrease in the collecting surface coverage value Θ_s. An expression for this decrease in Θ_s can be obtained as follows.

The total sputter erosion rate of 'fresh target metal atoms', R_m, can be written as

$$R_m = \left(\frac{J}{q}\right)\gamma_m(1 - \Theta_t)A_t$$

This sputter eroded metal gives rise to a uniform deposition rate R_m/A_s on the collecting surface. The amount of R_m that is deposited onto the compound fraction Θ_s of the collecting surface is

$$R_m\Theta_s$$

This term describes the rate of decrease in Θ_s due to deposition of unreacted metal atoms. For simplicity the terms R_m, $R_m\Theta_s$ and $R_m(1 - \Theta_s)$ are illustrated as arrows in figure A5.3.1.

Note that the complementary fraction $R_m(1 - \Theta_s)$ of sputtered metal deposited onto the metal fraction $(1 - \Theta_s)$ of the collecting area will not change the metal fraction at the collecting surface A_s since depositing metal onto metal does not change the composition at this position. The same holds for depositing sputtered compound material onto the fraction Θ_s at the collecting surface. Depositing compound material onto compound material does not change the composition of the area $A_s\Theta_s$.

Compound material can be formed by two different processes at the collecting surface. First the fresh metal fraction $(1 - \Theta_s)$ of the collecting surface may react with the reactive gas thereby forming compound material on this surface. The total rate of compound formation for this process is

$$\Gamma_r\alpha_s(1 - \Theta_s)A_s b$$

where α_s is the sticking coefficient for the reactive gas to the metal surface at the collecting surface and $(1 - \Theta_s)A_s$ is the metal part of the collecting surface.

The second process that may lead to compound formation is the deposition of sputtered compound material onto the metal fraction $(1 - \Theta_s)$ of the collecting area. The total sputter erosion rate R_c of compound material from the target is

$$R_c = \left(\frac{J}{q}\right)\gamma_c\Theta_t A_t$$

This sputtered compound material is uniformly distributed over the collecting area A_s. The part that is deposited onto the metal fraction $(1 - \Theta_s)$ of the collecting surface is

$$R_c(1 - \Theta_s)$$

This term defines the rate of increase in Θ_s due to deposition of sputtered compound material onto the fresh metal part of the collecting surface. For illustration also the terms R_c, $R_c\Theta_s$ and $R_c(1 - \Theta_s)$ are shown as arrows in figure A5.3.1.

At steady state the two terms that contribute to compound formation at the collecting surface must balance the term that describes the compound elimination at this surface. This defines the second basic balance equation

$$R_c(1 - \Theta_s) + \Gamma_r\alpha_s(1 - \Theta_s)A_s b = \left(\frac{R_m\Theta_s}{d}\right) \qquad \text{(A5.3.4)}$$

where d is a factor that defines the number of metal atoms in one compound molecule. The right-hand side of equation (A5.3.4) is divided by d to ensure that both sides are expressed in the same unit, i.e. compound molecules per time unit. When Θ_t has been calculated from equation (A5.3.2) it is possible to calculate Θ_s from equation (A5.3.4).

Gas kinetics

Normally the experimental equipment is not outfitted to monitor the partial pressure of the reactive gas during processing. The reactive gas is normally fed into the vacuum chamber via a mass flow controller. This is the reason why the supply of the reactive gas normally is used as an input parameter when a reactive sputtering process is described. In the model we would therefore want to find an expression for the supply of the reactive gas. This can be done in the following way.

As has been described above, the reactive gas will react with 'fresh metal surfaces' both at the target and at the collecting surfaces. Due to this effect reactive gas molecules will be gettered (extracted from the gas phase by forming solid compound molecules) at these surfaces. The gettering rate Q_t (number of molecules per unit time) at the target surface at a specific partial pressure of the reactive gas can be written as

$$Q_t = \Gamma_r \alpha_t (1 - \Theta_t) A_t \tag{A5.3.5}$$

The corresponding gettering Q_s of reactive gas at the collecting surface will be

$$Q_s = \Gamma_r \alpha_s (1 - \Theta_s) A_s \tag{A5.3.6}$$

At a partial pressure P_r of the reactive gas the vacuum pump of the system will pump out Q_p (number of molecules per unit time)

$$Q_p = k_1 P_r S \tag{A5.3.7}$$

where S is the pumping speed of the vacuum pump and k_1 is a factor converting the expression to the desired unit. Equations (A5.3.5)–(A5.3.7) determine the total consumption rate Q of reactive gas in the processing chamber. At steady state the contribution from equations (A5.3.5)–(A5.3.7) must balance the total input flow Q of reactive gas. This defines the third basic balance equation

$$Q = Q_t + Q_s + Q_p \tag{A5.3.8}$$

Since Γ_r, Θ_t and Θ_s can be calculated from equations (A5.3.1)–(A5.3.4) it is possible to also calculate Q from equation (A5.3.8) for all values of P_r. Equation (A5.3.8) states that to maintain steady state (constant partial pressure of the reactive gas) it is necessary to supply the reactive gas at a rate Q to the chamber to compensate for the total loss Q of this gas due to the three gettering mechanisms described by equations (A5.3.5)–(A5.3.7).

Calculations

Equations (A5.3.1)–(A5.3.8) uniquely describe the reactive sputtering process under the assumptions made. Since the gas supply of the reactive gas (Q) normally is the parameter used to vary the conditions during reactive sputtering processes we have chosen to use this parameter as an independent parameter in the graphical representation of our results. The simplest way to obtain results from equation (A5.3.1)–(A5.3.8) is to perform the calculations in the following sequence.

First select suitable values for the system parameters J, A_t, A_s, γ_m, γ_c, α_t, α_s, T, m, b and d.

- Assume a certain value for P_r
- Use equation (A5.3.1) to calculate the corresponding flux value Γ_r
- Calculate Θ_t from equation (A5.3.2)
- Calculate Θ_s from equation (A5.3.4)
- Calculate Q_t, Q_c, Q_p and Q from equations (A5.3.5)–(A5.3.8)
- Calculate R from equation (A5.3.3)
- Select a new value for P_r and repeat the calculation sequence, etc. In this way it is possible to generate a number of processing curves illustrating the relations between the different processing parameters. Any spread sheet program can be used to carry out these calculations. Below we show the results of such calculations.

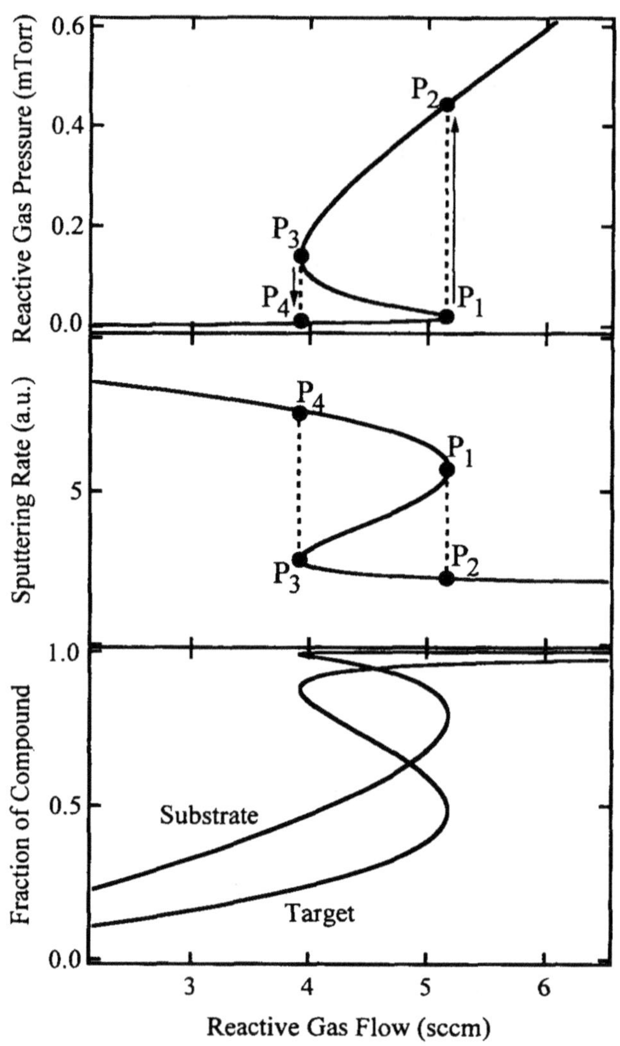

Figure A5.3.2. (Top) Simulation results of the reactive gas pressure versus flow in a standard reactive sputtering process. The points (P_1–P_4) are referred to in the text. (Middle) Simulation results of the sputtering rate versus reactive gas flow in a standard reactive sputtering process. The points (P_1–P_4) are referred to in the text. Same simulation parameters as in top part. (Bottom) Simulations results illustrating the different compositions of the target and the substrate surfaces as a function of the reactive gas flow. Same parameters as in top part.

The relation between the partial pressure and the supply of the reactive gas is shown in figure A5.3.2 (top). Note that each value of the partial pressure uniquely defines a corresponding value of the reactive gas supply Q. However, the same simple situation does not generally hold for each value of the reactive gas supply Q. In a certain region of the curve three different values of the partial pressure satisfy conditions resulting in identical values of the reactive gas supply. This region of the curve is called the hysteresis region. The reason for this can be explained in the following way.

Starting from zero flow, a small increase in the supply of the reactive gas will not significantly increase the partial pressure of the reactive gas. The reason for this is that most of the reactive gas supplied to the processing chamber is consumed by reactions with metal atoms at the surfaces A_t and A_s. Therefore only a small fraction of the reactive gas will be left over to build up a partial pressure of this gas. At a certain point P_1, however, the balance conditions for the partial pressure and the gettering at A_s and A_t will change drastically. As the partial pressure slowly increases, Θ_t will slowly increase causing less unreacted atoms to be sputtered and deposited onto the collecting surface. Since fewer unreacted atoms arrive at A_s, less reactive gas will be consumed in reactions with these atoms. Due to this effect an increasing fraction of the supplied reactive gas will contribute to building up an increase of the partial pressure of this gas. An increase of the partial pressure in turn will cause an even more effective compound formation at the target and subsequently even less pure metal sputtering. At P_1 this positive feedback effect will cause an avalanching increase of the partial pressure of the reactive gas. This avalanching effect will continue until the process reaches another stable point of operation. This new stable point of operation is denoted as P_2 in figure A5.3.2 (top). Increasing the supply of the reactive gas beyond this point will result in almost all the excess gas contributing to an increase of the partial pressure. No further reactive gas is needed to react with the sputtered metal since primarily almost only the metal compound is sputter eroded in this mode of operation.

At this point, decreasing the gas supply will initially result only in a decrease in the partial pressure. Passing the point P_2, however, will not result in a sharp decrease of the partial pressure. Since the latter is far too high to allow any substantial metal sputtering from the target, no significant increase in the consumption of reactive gas will take place at the collecting area A_s at the processing conditions corresponding to P_2. A further decrease of the supply of the reactive gas will ultimately result in reaching a point P_3 where 'insufficient' reactions with the gas at the target surface will result in an increase of unreacted metal atoms at this surface. At this point more unreacted target atoms will be sputtered and deposited onto the collecting area A_s. However, since A_s normally is much larger than A_t, the probability of reacting this metal at the surface A_s will be higher than at the target surface. Therefore gettering of the reactive gas will take place at A_s and subsequently less gas will remain to provide a partial pressure of the reactive gas. Therefore the partial pressure of the reactive gas will drop. This in turn will cause a further increase of sputtering of unreacted metal atoms that will cause an even larger gettering of the gas, etc. This is also an avalanching effect resulting in that the process avalanches from P_3 to the position P_4 on the processing curve.

From the above description it can be understood how the hysteresis effect appears during reactive sputtering. The hysteresis appears when increasing and decreasing the reactive gas flow without any further control of this gas. The process will thus follow the dotted lines in the figure. It is an unavoidable effect of the complex gettering mechanisms that occur during processing.

It should be noted, however, that if the processing chamber is equipped with some kind of a control unit capable of controlling the partial pressure in the chamber, it will be possible to follow the theoretical solid line in figure A5.3.2 (top). In fact this curve was calculated by first assuming a certain value of the partial pressure and thereafter calculating the necessary flow of the reactive gas. Experimentally this is done in the same manner. The control unit will automatically adjust the gas flow to obtain the desired partial pressure.

The corresponding curve for the sputtering rate R versus reactive gas flow, Q, is shown in figure A5.3.2 (middle). For the same reason as described above this curve also exhibits the hysteresis effect. The dotted line is the expected experimental curve when the partial pressure is not controlled. The solid curve corresponds to the same conditions as the solid curve in figure A5.3.2 (top).

In figure A5.3.2 (bottom) is shown the corresponding surface coverage curves Θ_t and Θ_s as a function of the reactive gas flow Q. It should be noted that Θ_s represents the composition of the deposited film and that Θ_t represents the so called 'target poisoning', i.e. the fraction of compound on the target surface. $\Theta_t = 0$ represents a clean metal target while $\Theta_t = 1$ represents target conditions where all metal atoms at the target surface have reacted with the reactive gas. In the case of reactive sputtering of Al in a mixture of oxygen and argon the sputter erosion rate from the target may differ by more than one order of magnitude between $\Theta_t = 0$ and $\Theta_t = 1$. It has been experimentally found, however, that it may be possible to operate the process at the high sputter rate mode as close as possible to the avalanche point (P_1) and still obtain a film coating having $\Theta_s = 1$ (Al_2O_3). This can easily be done in the partial pressure control mode if the feedback control has a fast enough response, but is much more delicate in the reactive gas supply mode since the risk of an avalanche runaway is obvious in this mode of operation close to this processing point.

The S-shape (solid lines) of the theoretical curves has been experimentally verified in systems having partial pressure control [3, 8]. In such controlled systems it is possible to make small increments of P_r

and obtain uniquely defined values of R, Q, Θ_t and Θ_s.

If the system is only supplied with a mass flow controller and has no system for pressure control the situation is somewhat different. In such systems only small increments in Q can be generated. At point P_1 in figure A5.3.2 (top), the system will respond to any further increase in Q by avalanching to the corresponding upper value (P_2) on the curve (dotted line). From here on the process will follow the solid line for a further increase in Q.

By decreasing the value of Q from a high level to smaller values the process will follow the solid line down to point P_3. A further decrease in the value of Q will result in that the process avalanches down to the corresponding lower value (P_4). From here on the process will follow the solid line for any further decrease in the value of Q.

The positions of the upward-avalanche and the downward-avalanche are clearly separated from each other. This separation defines the hysteresis width of the reactive sputtering process. This is a characteristic feature of most reactive sputtering processes.

It is possible to eliminate the hysteresis effect by increasing the pumping speed of the vacuum pump [9]. To illustrate this, a set of R versus Q curves are plotted in figure A5.3.3 for different pumping speeds of the vacuum system. The calculations clearly predict that the hysteresis is eliminated as the pumping speed is increased.

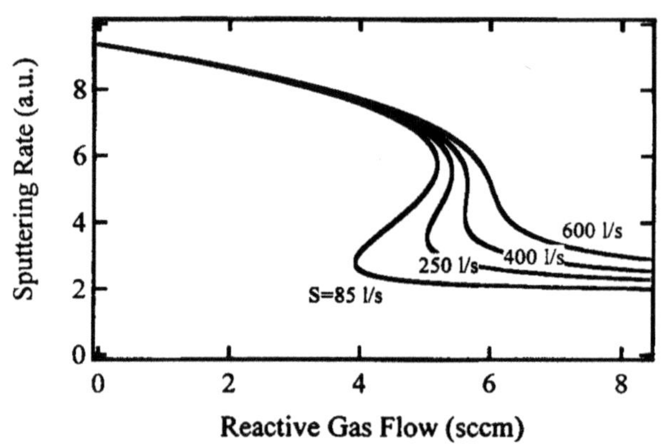

Figure A5.3.3. Simulation results illustrating the influence of the external pumping speed on the hysteresis width.

Another interesting processing behavior is illustrated in figure A5.3.4. In this figure, curves for R versus Q for different ion current densities J are plotted. It can be shown from the model outlined above that the dotted straight lines from the origin in this figure represent constant compositions (Θ_t and Θ_s) of the target and collecting surfaces [10]. This figure thus to some extent gives information on deposition rate as well as illustrating the effect of the composition of the deposited thin films. This figure also illustrates the aforementioned importance of applying the different parameters in a proper sequence. Applying 5.5 sccm of reactive gas flow and then a target current of 1.2 A will cause the process to move in the directions indicated by the arrows towards point P_A. If the target current is applied first, however, the process will follow the arrows to point P_B. Subsequently, it is possible to operate the process in both metallic and compound mode for the same reactive gas flow and target current depending on the sequence they are applied.

A5.3.2 REACTIVE CO-SPUTTERING

Sometimes it may be desirable to deposit a multi-element oxide, nitride or carbide, etc, by reactive sputtering. This can be realized by sputtering from several elemental metal targets (co-sputtering). At first sight this might seem like a simple solution for multi-elemental reactive sputter deposition. However, it is not possible to vary the conditions at the different targets independently. A power increase at one target in order to increase the deposition rate of this element will cause an increase in the gettering rate of reactive

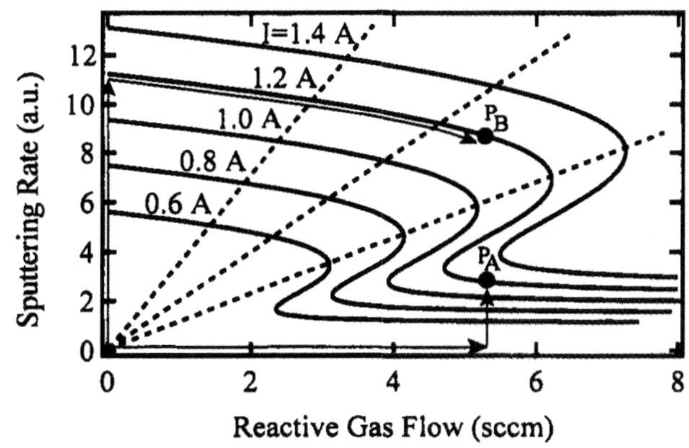

Figure A5.3.4. Simulation results of the sputtering rate versus reactive gas flow for different constant target currents. The dotted straight lines represent constant Θ_t and Θ_s.

gas molecules. This in turn will cause a decrease in the partial pressure of the reactive gas. This decrease will also affect the composition of the material sputtered from the other target, since both targets are in the same vacuum system and hence are exposed to identical partial pressure of the reactive gas.

It is quite straightforward, however, to apply the reactive sputtering model presented above to this case as well. Below we will outline the treatment of reactive sputtering from two separate elemental metal targets T_1 and T_2 (see figure A5.3.5(a).

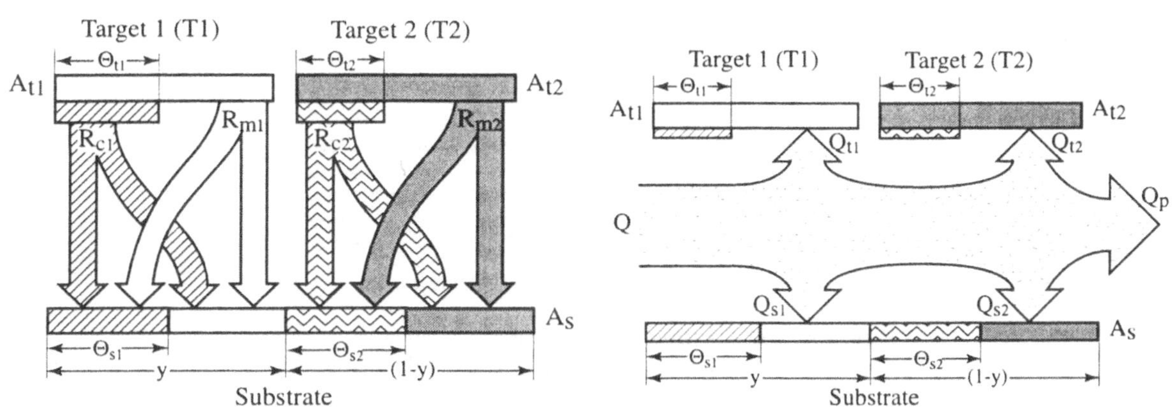

Figure A5.3.5. (a) Schematic of a two-target reactive sputtering system illustrating the flows of sputtered material. The expressions are referred to in the text. (b) An illustration of the reactive gas flux during reactive sputtering of two targets. The gas is either gettered at one of the targets ($Q_t = Q_{t1} + Q_{t2}$) or substrate ($Q_s = Q_{s1} + Q_{s2}$) metal surfaces or pumped out (Q_p).

Two balance equations for the individual targets (identical to the single target case derived above) can be derived by assuming that the partial pressure of the reactive gas is identical at the two targets.

The sputtered metal and metal-compound rates from T_1 are denoted R_{m1} and R_{c1} respectively. The corresponding outsputtered materials from T_2 are denoted R_{m2} and R_{c2} respectively. All the outsputtered material is assumed to be evenly collected at the collecting surface A_s. We then define a fractional 'composition' y of the deposited material at the collecting surface A_s as

$$y = \frac{R_{m1} + d_1 R_{c1}}{(R_{m1} + d_1 R_{c1}) + (R_{m2} + d_2 R_{c2})}$$

where d_1 and d_2 define the number of metal atoms in the corresponding compound molecules. For simplicity and similarly to the arguments above we assume that the sputtered material from target T_1 will be collected at the fraction y of the collecting surface A_s and that the sputtered material from target T_2 is collected at the complementary fraction $(1 - y)$ of the collecting surface A_s. Mathematically we treat this situation as two separate collecting surfaces for materials coming from targets T_1 and T_2 respectively. The problem of describing a dual target reactive sputtering process has thus been simplified to consist of two single element reactive sputter deposition processes linked together via the condition that they both must have the same partial pressure of the reactive gas. Note that the fractional 'composition' y will change its value according to the processing conditions. In figure A5.3.5(a) we show a schematic outline representing the flux of sputtered material for a reactive co-sputtering system with two separate targets. In figure A5.3.5(b) the corresponding flow of the reactive gas is also shown. Note that there are four different areas where the gas may react with 'fresh' metal surfaces. One at each target and one at the metal M_1 fraction yA_s and one at the metal M_2 fraction $(1 - y)A_s$ of the collecting surface. The mathematical treatment of the target surfaces and the collecting surfaces yA_s and $(1 - y)A_s$ can easily be dealt with analogous to the treatment of two single target reactive sputtering processes joined together by sharing identical partial pressure of the reactive gas [4].

The behavior of a reactive co-sputtering process having two different targets can be predicted by applying the mathematical treatment outlined above. Results from such calculations are shown below.

In figure A5.3.6 are shown the R versus Q curves for three different processing conditions. Included in this figure are first the R versus Q curves for each target calculated under the assumption that no sputtering takes place from the other target. Due to the different reactivity of the two target materials the hysteresis width and the position of the hysteresis may be quite separated. However, when both targets are in operation simultaneously, the reactive co-sputtering process will only generate one hysteresis region.

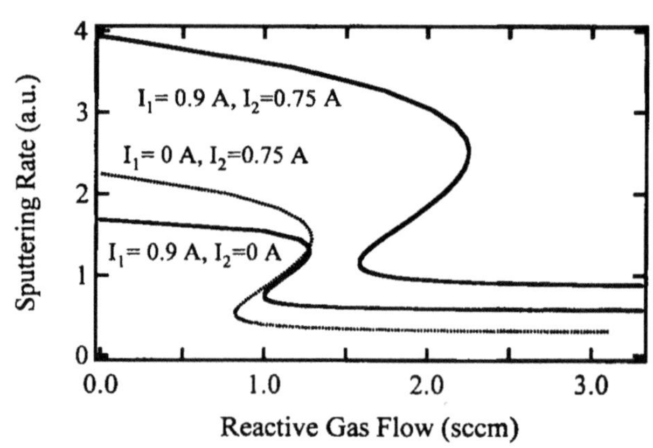

Figure A5.3.6. Simulation of a dual target reactive sputtering system for different combinations of the target currents.

The fractional 'composition' y in the deposited film may change dramatically. This is shown in figure A5.3.7. In this case y exhibits a loop in the hysteresis region. This illustrates an extreme processing behavior that may occur during reactive co-sputtering. This variation in composition is due to the fact that the sputtering rates from the individual targets may change quite differently depending on the reactivity of the target atoms to the reactive gas as the flow is varied. This behavior has been experimentally observed for reactive co-sputtering of Ti and V in nitrogen gas [11].

A5.3.3 REACTIVE SPUTTERING FROM AN ALLOY TARGET

To avoid the problem of controlling the composition of the deposited film in reactive co-sputtering it may be advantageous to carry out reactive sputtering from an alloy target. The mathematical description of this deposition process will almost be identical to the treatment of the reactive co-sputtering. However, we have to calculate the alloy target surface composition since this composition, owing to the so called preferential sputtering effects [6] will differ from the bulk composition of the target. This deviation from the bulk composition results in a pressure dependent surface composition. This is illustrated in figure A5.3.8 together with all material flows. For a two element alloy target with a bulk composition of $M1_xM2_{(1-x)}$ this can be treated in the following way.

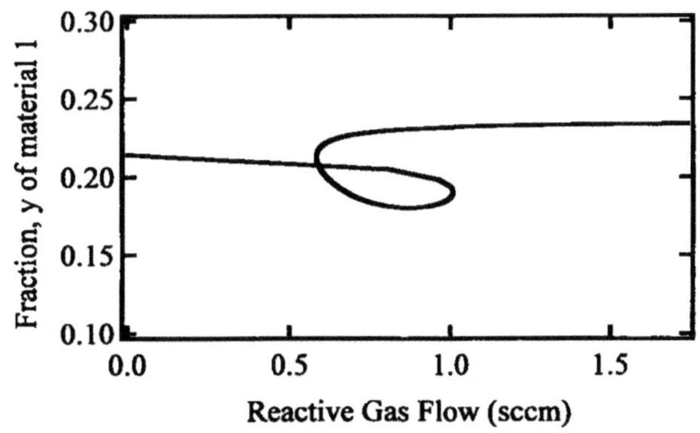

Figure A5.3.7. The relative amount of material from target 1 in the deposited film versus reactive gas flow according to simulations of a two target system.

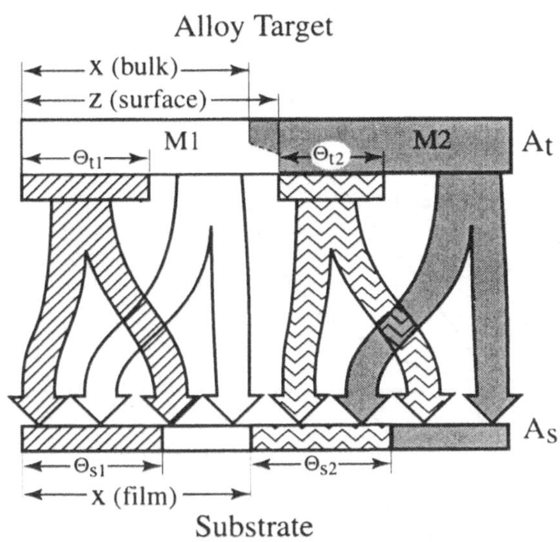

Figure A5.3.8. A schematic of an alloy target reactive sputtering system illustrating the flows of sputtered material. The expressions are referred to in the text. Note that the surface concentration z of the target differs from the target bulk concentration x.

As in the previous treatments we separate the target materials at the surface in two separate regions assuming that the 'composition' of the surface during processing will be $M1_zM2_{(1-z)}$ where M1 represents the first metal and M2 represents the second one irrespective of whether they are in pure metal or compound state. The total number of metal atoms of type M1 that are sputter eroded from the target surface region zA_t will be

$$R_1 = \left(\frac{J}{q}\right) zA_t(\gamma_{m1}(1-\Theta_{t1}) + d_1\gamma_{c1}\Theta_{t1}) \tag{A5.3.9}$$

and the corresponding expression for the metal 2 surface region $(1-z)A_t$ will be

$$R_2 = \left(\frac{J}{q}\right)(1-z)A_t(\gamma_{m2}(1-\Theta_{t2}) + d_2\gamma_{c2}\Theta_{t2}) \tag{A5.3.10}$$

The ratio between the sputter erosion rates R_1 and R_2 must correspond to the bulk composition of the target. The following relation must therefore hold

$$\frac{R_1}{R_2} = \frac{x}{1-x} \tag{A5.3.11}$$

The alloy target can thus be treated like the reactive co-sputtering having two separate targets one having a surface area zA_t (metal 1) and the other a surface area $(1-z)A_t$ (metal 2). z will vary and has to be calculated for each value of the partial pressure of the reactive gas. Θ_{t1} and Θ_{t2} can be calculated by applying the same balance description for the separate target surfaces as for the single target case. By inserting the values of Θ_{t1} and Θ_{t2} in equation (A5.3.11), z can also be calculated.

Equation (A5.3.11) ensures that the model correctly incorporates the obvious steady-state conservation requirement that the relative concentrations of M1 and M2 type of atoms are identical to their relative bulk concentrations. It should be noted, however, that this condition does not imply that both metals react at the same rate with the reactive gas. Calculations based on the arguments described above are shown in figure A5.3.9 where the film composition is plotted versus reactive gas flow. The reactive gas has arbitrarily been assumed to have a sticking coefficient to M2 of 0.05 and the value 0.3 for M1. It is clearly seen that one may expect that metal 1 forms a compound much easier than metal 2. Consequently, operating close to the avalanche point P_1 may not necessarily yield films in which both metal species are fully reacted, as is often the case of the single element reactive sputtering process. To ensure fully compound formation one may have to increase the supply of the reactive gas far beyond the avalanche point and thus carry out the deposition process in the low sputtering rate mode.

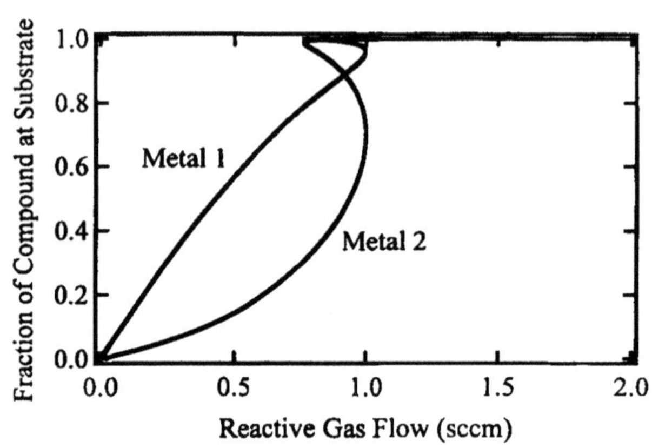

Figure A5.3.9. Fractions of the different metals that have formed a compound versus reactive gas flow according to simulations of reactive sputtering from an alloy target.

A5.3.4 REACTIVE SPUTTERING INCLUDING TWO REACTIVE GASES.

By adding two reactive gases to a sputter deposition process it is possible to form e.g. oxy-nitrides and carbon-nitrides, etc. At first sight this might look as not too complicated a process. However, as has already been pointed out, the reactive sputtering process is extremely sensitive to the partial pressure of the reactive gas. Using two reactive gases will add an extra very sensitive parameter to the process. For an elemental target that forms separate compounds with the two reactive gases the conditions at the target are illustrated in figure A5.3.10.

We may define two surface coverage coefficients Θ_{t1} and Θ_{t2} representing the fractions of the surfaces that are covered by the compound formed by reactions with gas 1 and gas 2 respectively. The arguments to obtain balance equations will be very similar to the previous cases.

At the target, the compound formation rate by gas 1 has to be equal to the sputter removal rate of compound 1 from the fractional area Θ_{t1} of A_t. This yields the following balance equation for the compound 1 part of the target:

$$\Gamma_{r1}\alpha_{t1}\,[1-(\Theta_{t1}+\Theta_{t2})]\,A_t b_1 = \left(\frac{J}{q}\right)\gamma_{c1}\Theta_{t1}A_t \tag{A5.3.12}$$

where the indices and the symbols are defined analogous to previous definitions. A similar balance equation can be defined for the compound 2 part of the target.

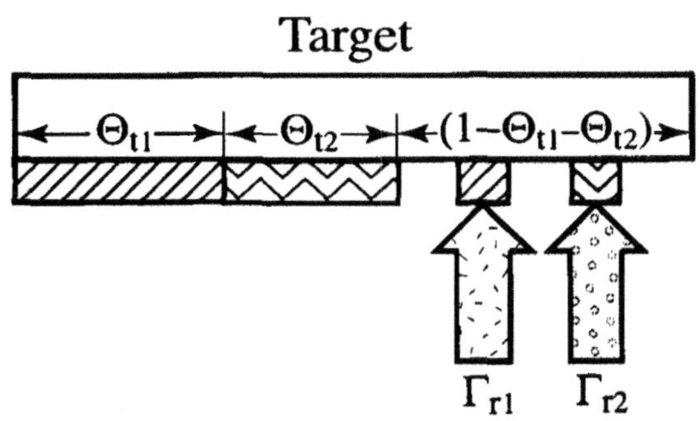

Figure A5.3.10. A schematic illustrating how the target is assumed to be divided into three parts for reactive sputtering using two gases. One pure metal part and one part for each compound.

At the collecting surface one may also define three separate regions, the unreacted metal region, the compound 1 region and the compound 2 region. By similar arguments as described above it is also possible to define balance equations for this area. Due to the similarities with the previous treatments, however, we do not include the mathematical details in this presentation. The interested reader is referred to [5] for further details. We will, however, make some remarks concerning this process.

It is recalled that an increase in the partial pressure of a reactive gas causes an increase in the fraction of the target that is covered by the compound. This in turn causes a decrease in the sputter erosion rate of unreacted metal atoms causing a decrease in the gas gettering which in return reduces gas consumption that further increases the partial pressure and so on.

In the two gas reactive sputtering process, any of the two gases may start this avalanche. In fact, it may happen that one of the gases may poison the target to such an extent that adding a small amount of the second gas will trigger an avalanche. This effect is illustrated in figure A5.3.11. This figure shows the results of the simulations during the conditions that a constant supply of one reactive gas (1) is supplied to the processing chamber. This supply level is not large enough to force the process to pass the avalanche point from high sputter rate mode to low sputter rate mode. However, by adding a supply of the second gas (2) the process is forced to pass the avalanche point. From the figure it can be seen that this may well result in that an increase in the supply of gas 2 causes a larger increase in compound 1 formation than compound 2 opposite to what one might expect. An increase in the supply of gas 2 causes a larger increase in the partial pressure of gas 1 than of gas 2. Due to this effect formation of compound 1 will be favored.

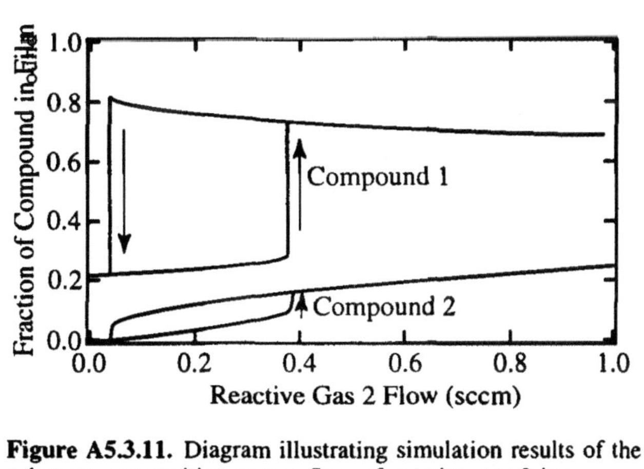

Figure A5.3.11. Diagram illustrating simulation results of the substrate composition versus flow of reactive gas 2 in a two-reactive-gas sputtering system. The flow of gas 1 is assumed to be constant.

The above effect serves to illustrate that it may be quite a delicate matter to control the partial pressure of the two gases during reactive sputtering with two reactive gases. Since the supply of one gas significantly may influence the partial pressure of the other there exists no simple straightforward way of applying process control in this case.

A5.3.5 CONCLUSIONS

The above presentation serves to illustrate that by quite simple physical arguments it is possible to set up mathematical models for different reactive sputtering processes. Despite all simplifications made to obtain simple and manageable models the results predicted by these models fit remarkably well to most experimental results published in the literature. Hopefully this may encourage the reader to add his/her own improvements to the model by modifying some of the equations to obtain better accuracy.

Computer power has reached such a level that the reason for making crude approximations to obtain simple mathematical expressions may no longer be necessary. Therefore more detailed information of the reactive sputtering process may now be included to form more advanced models of the reactive sputtering process. We will point out a few new ways of expanding the computer simulations of the reactive sputter deposition process.

It is well known that the sputtering yield for an ion impinging on a surface is strongly dependent on e.g. the matrix and composition of the surface layer. However, there exist computer simulation codes that with reasonable accuracy predict these variations [7]. Since the composition of the target surface varies for different processing conditions it should be valuable to include the variation in the sputtering yield due to this effect in a reactive sputtering model.

The transport of sputtered material from the target to the substrate (and chamber walls) is quite more complicated than assumed in the models above. There exist, however, Monte Carlo based simulation programs [12] that calculate scattering due to gas collisions and also the flux distribution of the sputtered material in the vacuum chamber. Furthermore, the race-track (the limited area on the target surface from which material is sputter eroded) of a magnetron target indicates that the ion current density does vary significantly on the target surface resulting in non-uniform sputter deposition fluxes at various receiving surfaces. Due to this effect material is sputter eroded inhomogenously from the target surface. Such effects may also be included in the model. By considering a sufficient number of collecting surfaces much higher precision in the calculations of the gettering of the reactive gas may be obtained.

The current through the target consists of the ion current and the electron current originating from secondary electrons created by the energetic ions impacting onto the target surface. This electron current may contribute to a substantial part of the total current. Moreover, the secondary electron yield depends strongly on the surface composition. Due to this effect a constant current in the external electric circuit does not necessary imply a constant ion current during the reactive sputtering process. To increase the precision in the reactive sputtering modeling work the effect of the variation in the secondary electron yield should also be included.

The list of 'second-order effects' can of course be very long. One has to justify the amount of work put into the model against the expected value of the improvements of the calculations. In most practical applications it is quite satisfactory, if possible, to obtain a good solid basic understanding of the processing behavior. The 'fine tuning' of the process will be done on-site in any case. However, accurate processing models are necessary in the work of obtaining adequate process control.

REFERENCES

[1] Westwood W D 1989 *Handbook of Plasma Processing Technology* (Park Ridge, NJ: Noyes Publications) pp 233–59

[2] Berg S, Blom H-O, Larsson T and Nender C 1987 *J. Vac. Sci Technol.* A **5** 202–7

[3] Sekiguchi H, Kanzawa A, Imai T and Honda T 1994 *J. Vac. Sci. Technol.* A **12** 3176–9

[4] Moradi M, Nender C, Berg S, Blom H-O, Belkind A and Orban Z 1991 *J. Vac. Sci. Technol.* A **9** 619–24

[5] Carlsson P, Nender C, Barankova H and Berg S 1993 *J. Vac. Sci. Technol.* A **11** 1534–39

[6] Behrisch R, Sigmund P, Robinson M T, Andersen H H, Bay H L and Roosendaal H E 1981 *Sputtering by Particle Bombardment* I (Berlin: Springer)
[7] Nender C, Katardjiev I V, Biersack J P, Berg S and Barklund A M 1994 *Radiation Effects and Defects in Solids* **281** 130–31
[8] Moradi M, Nender C, Blom H-O and Berg S 1990 *Vacuum* **41** 1974–6
[9] Larsson T, Blom H-O, Nender C and Berg S 1988 *J.Vac. Sci. Technol.* A **6** 1832–6
[10] Berg S, Larsson T, Nender C and Blom H-O 1988 *J. Appl. Phys.* **63** 887–91
[11] Nyberg T, Skytt P, Gålnander B, Nender C, Nordgren J and Berg S 1997 *J.Vac. Sci. Technol.* A **15** 145–7
[12] Alberta Microelectronic Center, Edmonton, Alberta, Canada 1997

Appendix A

List of Contributors

J R Abelson (X2.2)
University of Illinois at Urbana-Champaign,
1-109 Engineering Science Building,
1101 West Springfield Avenue,
Urbana, IL 61801,
USA

C R Abernathy (A2.1)
Department of Material Science,
University of Florida,
Gainesville, FL 32611,
USA

John Agostinelli (X4.2)
Eastman Kodak Research Laboratories,
MC 02011,
Rochester, NY 14650-2011,
USA

Jinho Ahn (X2.16)
Department of Materials Engineering,
Hanyang University,
Seoul 133-791,
Korea

Harry A Atwater (X2.15)
Thomas J Watson Laboratory of Applied Physics,
California Institute of Technology,
Pasadena, CA 91125,
USA

O Auciello (X3.4)
Department of Materials Science and Engineering,
North Carolina State University,
Raleigh, NC 27695-7919,
USA

Jacob Azoulay (X6.0)
Center for Technological Education Holon,
52 Golomb Street,
PO Box 305,
Holon 58102,
Israel

Scott A Barnett (A2.0)
Materials Science Department,
Northwestern University,
2225 Sheridan Road,
Evanston, IL 60202,
USA

T Baron (X2.4)
CEA-DRFMC/SP2M,
Laboratoire de Physique des Semiconducteurs,
17 Avenue des Martyrs,
38054 Grenoble Cedex 9,
France

John C Bean (X2.6)
Room 1C-326,
AT&T Bell Labs,
Murray Hill, NJ 07974,
USA

Rainer Beccard (X2.4, X4.3, X6.2)
AIXTRON Semiconductor Technologies GmbH,
Kackerstrasse 15–17,
D-52072 Aachen,
Germany

S Berg (A5.3)
Solid State Electronics,
The Angstrom Laboratory,
Uppsala University,
Box 534,
S-751 21 Uppsala,
Sweden

Paul R Berger (D2.4)
University of Delaware,
Department of Electrical Engineering,
Newark, DE 19716,
USA

Graciela B Blanchet (X7.1)
E.I. DuPont de Nemours,
Experimental Station,
PO Box 80356,
Wilmington, DE 19898-0356,
USA

H-O Blom (A5.3)
Solid State Electronics,
The Angstrom University,
Uppsala University,
Box 534,
S-751 21 Uppsala,
Sweden

A Botchkarev (X2.10)
Coordinated Science Laboratory and Material Research Laboratory,
University of Illinois at Urbana-Champaign,
104 South Goodwin Avenue,
Urbana, IL 61801,
USA

T R Bramblett (X2.6, X2.7)
University of Illinois at Urbana-Champaign,
1101 West Springfield Avenue,
Urbana, IL 61801,
USA

Robert A Breun (D3.1)
University of Wisconsin,
Department of Nuclear Engineering,
1500 Johnson Dr.,
Madison, WI 53706-1687,
USA

Charles Bruker (X5.1, X5.2)
Eastman Kodak Research Laboratories,
MC 02017,
Rochester, NY 14650-2017,
USA

Ken Cadien (X2.0)
Intel Corporation, SC11-104,
2200 Mission College Blvd,
Santa Clara, CA 95052-8119,
USA

Robert C Cammarata (F2)
Department of Materials Science and Engineering
The Johns Hopkins University
Baltimore, MD 21218
USA

Stephen A Campbell (X3.5)
University of Minnesota,
Department of Chemistry,
Kolthoff and Smith Halls,
Minneapolis, MN 55455-0431,
USA

P F Carcia (X5.3)
E.I. DuPont de Nemours,
Experimental Station,
PO Box 80356,
Wilmington, DE 19898-0356,
USA

James S Chan (X2.11)
Department of Electrical Engineering & Computer Sciences,
University of California at Berkeley,
CA 94720,
USA

Subhas Chandra (X3.2)
National Physical Laboratory,
Dr K S Krishnan Road,
New Delhi-110012,
India

Sheng-Liang Chang (D1.1)
Department of Chemistry and Ames Laboratory,
Iowa State University,
Ames, IA 50011,
USA

Dilip Chatterjee (X4.1, X4.2)
Eastman Kodak Research Laboratories,
MC 02011,
Rochester, NY 14650-2011,
USA

Nathan Cheung (X2.11)
Department of Electrical Engineering & Computer Sciences,
University of California at Berkeley,
CA 94720,
USA

T H Chiu (A2.3)
AT&T Bell Laboratories,
Holmdel, NJ 07733,
USA

V Chivukula (X4.0)
Northern Telecom Limited,
Nepean, Ontario,
Canada

Alice F Chow (X3.4)
Department of Materials Science and Engineering,
North Carolina State University,
Raleigh, NC 27695-7919,
USA

John W Coburn (C1.0)
6122 Franciscan Way,
San Jose, CA 95120-4416,
USA

R W Collins (D0)
The Pennsylvania State University,
Materials Research Laboratory,
University Park, PA 16802,
USA

Catherine M Cotell (E2.1)
Naval Research Laboratory, code 6670,
4555 Overlook Avenue, SW,
Washington, DC 20375-5345,
USA

John D Cressler (F6.0:2)
Department of Electrical Engineering
421 Broun Hall
Auburn University
Auburn, AL 36849-5201
USA

C Daly (D4.0)
Northeastern University,
Department of Physics,
110 Forsyth St.,
Boston, MA 02115,
USA

David F Dawson-Elli (X6.1)
Corning Inc,
SP-FR-3-1,
Corning, NY 14831,
USA

V M Donnelly (D2.2)
AT&T Laboratories,
Materials Science and Engineering Research Division,
Murray Hill, NJ 07974,
USA

Bernard Drévillon (D2.1)
Laboratoire de Physique des Interface et des Couches Minces (UPR 258 du CNRS),
Ecole Polytechnique,
91128 Palaiseau,
France

Russell D Dupuis (B1.1, X2.10, X2.13)
Microelectronics Research Center,
University of Texas at Austin,
MER 1.606 D/R9900,
Austin, TX 78712-1100,
USA

Robert W Eason (X3.4)
Department of Physics and Optoelectronics Research Centre,
University of Southampton,
Southampton SO17 1BJ,
UK

Grant O Este (A5.2)
7 Mohawk Crescent,
Nepean, ON,
K2H 7G7,
Canada

Dean Face (X6.2)
E.I. DuPont de Nemours,
Experimental Station,
PO Box 80304,
Wilmington, DE 19898-0304,
USA

Ian T Ferguson (A2.0)
Materials Research Center,
Northwestern University,
2225 Sheridan Road,
Evanston, IL 60202,
USA

A Figueras (X6.2)
Institut de Ciencia de Materials de Barcelona,
Campus de la UAB,
08193 Bellaterra,
Catalunya,
Spain

Tracy Fu (X2.11)
Department of Electrical Engineering & Computer Sciences,
University of California at Berkeley,
CA 94720,
USA

Eiji Fujii (X3.3)
Central Research Laboratories,
Matsushita Electric Industrial Co. Ltd,
Hikasidai, Seikacho,
Kyoto 619-02,
Japan

Eric E Fullerton (F4)
Argonne National Laboratory
Materials Science Division - Bldg 223
9700 South Cass Avenue
Argonne, IL 60439
USA

Bernard Gallois (D2.5)
Stevens Institute of Technology,
Department of Materials Science & Engineering,
Castle Point on the Hudson,
Hoboken, NJ 07030,
USA

Jorge Roberto Vargas Garcia (X7.3)
Institute for Materials Research,
Tohoku University,
2-1-1 Katahira, Aoba-ku,
Sendai 980-77,
Japan

L J Gerenser (E3.1)
Eastman Kodak Company,
Building 82,
Rochester, NY 14650-2132,
USA

Devinder S Gill (X3.4)
Department of Physics and Optoelectronics
Research Centre,
University of Southampton,
Southampton SO17 1BJ,
UK

David C Gilmer (X3.5)
University of Minnesota,
Department of Chemistry,
Kolthoff and Smith Halls,
Minneapolis, MN 55455-0431,
USA

Wayne L Gladfelter (X3.5)
University of Minnesota,
Department of Chemistry,
Kolthoff and Smith Halls,
Minneapolis, MN 55455-0431,
USA

David Glocker (D4.1, X3.1)
Eastman Kodak Research Laboratories,
Bldg 81, MC 02022,
Rochester, NY 14650-20,
USA

Takashi Goto (X7.3)
Institute for Materials Research,
Tohoku University,
2-1-1 Katahira, Aoba-ku,
Sendai 980-77,
Japan

T M Graettinger (X3.4)
Department of Materials Science and Engineering,
North Carolina State University,
Raleigh, NC 27695-7919,
USA

Michael E Graham (X1.1, X1.2, X1.3, X1.4, X1.5, X1.6)
BIRL,
1801 Maple Avenue,
Evanston, IL 60201-3135,
USA

E B Graper (A1.0, A1.1, A1.2, A1.3)
5960 Mandarin Avenue,
Goleta, CA 93117,
USA

S M Green (X6.2)
Neocera, Inc.,
335 Paint Branch Drive,
College Park, MD 20742,
USA

J E Greene (X2.6, X2.7)
University of Illinois at Urbana-Champaign,
1101 West Springfield Avenue,
Urbana, IL 61801,
USA

Henry J Gysling (X4.2)
Photoscience Research Division,
Imaging Research & Advance Development,
Eastman Kodak Company,
Rochester, NY 14650-2109,
USA

George Hadjipanayis (X5.0)
Department of Physics and Astronomy,
University of Delaware,
Newark, DE 19716,
USA

K S Harshavardhan (X6.2)
Neocera, Inc.,
335 Paint Branch Drive,
College Park, MD 20742,
USA

Gang He (X2.15)
Thomas J Watson Laboratory of Applied Physics,
California Institute of Technology,
Pasadena, CA 91125,
USA

Noah Hershkowitz (D3.0)
University of Wisconsin,
Department of Nuclear Engineering and Engineering Physics,
1500 Johnson Drive,
Madison, WI 53706-1687,
USA

H Higuma (X6.4)
Materials and Electronic Devices Laboratory,
Mitsubishi Electric Corp,
Sagamihara,
Kanagawa 229,
Japan

B J Hinds (X6.3)
Department of Chemistry,
Northwestern University,
2145 Sheridan Road,
Evanston, IL 60208-3113,
USA

Toshio Hirai (X7.3)
Institute for Materials Research,
Tohoku University,
2-1-1 Katahira, Aoba-ku,
Sendai 980-77,
Japan

William Holstein (X6.3)
E.I. DuPont de Nemours,
Experimental Station,
PO Box 80356,
Wilmington, DE 19898-0356,
USA

Wen L Hsu (X1.8)
Sandia National Laboratories,
Advanced Materials Research Department,
MS 9162, PO Box 969,
Livermore, CA 94551-0969,
USA

F Y Huang (F6.1.1)
Electrical Engineering Department
University of California, Los Angeles
Los Angeles, CA 90095-1594
USA

Graham K Hubler (E2.2)
Naval Research Laboratory, code 66670,
Washington, DC 20375,
USA

Stuart J C Irvine (B1.3)
North East Wales Institute,
Plas Loch, Mold Road,
Wrexham,
Clwyd LL11 2AW,
UK

Tadatugu Itoh (A3.3, X2.1)
School of Engineering,
Kobushikan University,
Setagaya 4-28-1, Setagaya-ku,
Tokyo 154,
Japan

B Jalali (F6.1.1)
Electrical Engineering Department
University of California, Los Angeles
Los Angeles, CA 90095-1594
USA

Frank Jansen (B1.2)
BOC Coating Technology,
PO Box 4105,
4020 Pike Lane,
Concord, CA 94524,
USA

David E Jesson (F1)
Solid State Division
Oak Ridge National Laboratory
Oak Ridge, TN 37831-6030
USA

Q X Jia (X3.6)
Los Alamos National Laboratory,
Superconductivity Technology Center,
M/S K763,
Los Alamos, NM 87545,
USA

Bruce A Joyce (D1.0)
Interdisciplinary Research Centre for Semiconductor Materials,
The Blackett Laboratory,
Imperial College,
Prince Consort Road,
London SW7 2BZ,
UK

M N Kamalasanan (X3.2)
National Physical Laboratory,
Dr K S Krishnan Road,
New Delhi-110012,
India

P A Kammer (A4.1, X1.9, X1.10, X1.11)
Castolin-Eutectic,
Schweissmaterials ag,
Postfach 1020,
1001 Lausanne,
Switzerland

Ron Kaspi (D1.2)
Wright Laboratory,
Solid State Electronics Directorate, WL/ELRA,
Wright-Patterson AFB, OH 45433-7323,
USA

A I Kingon (X3.4)
Department of Materials Science and Engineering,
North Carolina State University,
Raleigh, NC 27695-7919,
USA

Otto Knotek (X1.0)
Institut für Werkstoffkunde,
Rhein Westfälische Technische Hochschule,
Aachen Templergraben 55,
D-52056 Aachen,
Germany

James Kolodzey (X2.8, X2.13)
Department of Electrical Engineering,
University of Delaware,
Newark, DE 19716,
USA

J Krim (D4.0)
Northeastern University,
Department of Physics,
110 Forsyth St.,
Boston, MA 02115,
USA

N Deepak Kumar (X3.2)
National Physical Laboratory,
Dr K S Krishnan Road,
New Delhi-110012,
India

D N Lambeth (X5.0)
Data Storage Systems Center,
Department of Electrical and Computer Engineering,
Carnegie Mellon University,
Pittsburgh, PA 15213,
USA

Scott L Lantz (X2.9)
M/S FB7-57,
4100 Sara Road,
Rio Rancho, NM 87124,
USA

Keith O Legg (X1.1, X1.2, X1.3, X1.4, X1.5, X1.6)
BIRL,
1801 Maple Avenue,
Evanston, IL 60201-3135,
USA

Gregor Lengeling (X2.4, X4.3, X6.2)
AIXTRON Semiconductor Technologies GmbH,
Kackerstrasse 15–17,
D-52072 Aachen,
Germany

Zhuang Li (X4.3)
Materials Science Division,
Argonne National Laboratory,
Argonne, IL 60439,
USA

D J Lichtenwalner (X3.4)
Department of Materials Science and Engineering,
North Carolina State University,
Raleigh, NC 27695-7919,
USA

E M Liston (E3.0)
42 Peninsula Road,
Belvedere, CA 94920,
USA

Chih-shun Lu (D3.3)
Intelligent Sensor Technology, Inc.,
1012A Linda Vista Ave.,
Mountain View, CA 94043,
USA

Q J Lu (X2.6, X2.7)
University of Illinois at Urbana-Champaign,
1101 West Springfield Avenue,
Urbana, IL 61801,
USA

T J Marks (X6.3)
Department of Chemistry,
Northwestern University,
2145 Sheridan Road,
Evanston, IL 60208-3113,
USA

P J Martin (A1.4)
CSIRO Division of Applied Physics,
Sydney,
Australia 2070

L Martinu (E3.0)
Groupe des Couches Minces &
Department of Engineering Physics,
Ecole Polytechnique,
Box 6079, Station 'Centre-Ville',
Montreal, Quebec H3C 3A7,
Canada

Donald Mattox (E1.0)
440 Live Oak Loop,
Albuquerque, NM 87122,
USA

Kiyoshi Miyake (A3.4)
Power & Industrial Systems
R & D Division,
Hitachi Ltd, 1-1-1 Kokubu-cho,
Hitachi-shi, Ibaraki 316,
Japan

Akiharu Morimoto (A1.5)
Department of Electrical and Computer
Engineering,
Faculty of Technology, Kanazawa University,
Kanazawa 920,
Japan

Hadis Morkoç (A2.4, X2.1)
Coordinated Science Laboratory,
University of Illinois at Urbana-Champaign,
1101 West Springfield Avenue,
Urbana, IL 61801,
USA

J Myoung (X2.10)
Coordinated Science Laboratory and Material
Research Laboratory,
University of Illinois at Urbana-Champaign,
104 South Goodwin Avenue,
Urbana, IL 61801,
USA

T Nabatame (X6.3)
Superconductivity Research Laboratory,
Nagoya Division, ISTEC,
2-4-1 Matsuno, Atsuta-ku,
Nagoya 456,
Japan

Michael Nastasi (E2.2)
Los Alamos National Laboratory,
Center for Materials Science,
CMS, Mail Stop K765,
Los Alamos, NM 87545,
USA

C Nender (A5.3)
Solid State Electronics,
The Angstrom Laboratory,
Uppsala University,
Box 534,
S-751 21 Uppsala,
Sweden

Nate Newman (X2.11)
Lawrence Berkeley Laboratory,
1 Cyclotron Road,
University of California at Berkeley,
CA 94720,
USA

T Nyberg (A5.3)
Solid State Electronics,
The Angstrom Laboratory,
Uppsala University,
Box 534,
S-751 21 Uppsala,
Sweden

Chris Palmstrøm (X2.12)
University of Minnesota,
Dept of Chemical Engineering and Materials
Science,
151 Amundson Hall,
Minneapolis, MN 55455-0132,
USA

Alan S Penfold (A3.1, A3.2)
R.R. #2, Box 72,
Cornish, NH 03745,
USA

A Piqué (X6.2)
Neocera, Inc.,
335 Paint Branch Drive,
College Park, MD 20742,
USA

R C Powell (X2.10)
Solar Cells Inc,
1702 North Westwood Avenue,
Toledo, OH 43606,
USA

Roger W Pryor (X1.7)
Wayne State University,
666 West Hancock St.,
Detroit, MI 48202,
USA

Yi Qu (X3.2)
Institute of Energy Conservation,
University of Delaware,
Department of Electrical Engineering,
Newark, DE 19716,
USA

Donald J Rej (E2.3)
Los Alamos National Laboratory,
Mail Stop D434,
Los Alamos, NM 87545,
USA

Z F Ren (X6.3)
Superconductive Materials Laboratory,
Acheson Hall,
State University of New York at Buffalo,
Buffalo, NY 14214,
USA

Gary Y Robinson (A2.2)
Department of Electrical Engineering,
Colorado State University,
Fort Collins, CO 80523,
USA

A Rockett (X2.5)
University of Illinois Urbana-Champaign,
Engineering Sciences Building,
1101 W. Springfield Avenue,
Urbana, IL 61801,
USA

Mike Rubin (X2.11)
Lawrence Berkeley Laboratory,
1 Cyclotron Road,
University of California at Berkeley,
CA 94720,
USA

Paul J Rudnik (X1.1, X1.2, X1.3, X1.4, X1.5, X1.6)
BIRL,
1801 Maple Avenue,
Evanston, IL 60201-3135,
USA

Lourdes Salamanca-Riba (F6.2)
Materials and Nuclear Engineering Department,
University of Maryland,
College Park, MD 20742-2115,
USA

A Salvador (X2.10)
Coordinated Science Laboratory and Material
Research Laboratory,
University of Illinois at Urbana-Champaign,
104 South Goodwin Avenue,
Urbana, IL 61801,
USA

K Saminadayar (X2.4)
CEA-DRFMC/SP2M,
Laboratoire de Physique des Semiconducteurs,
17 avenue des Martyrs,
38054 Grenoble Cedex 9, France

F Sandiumenge (X6.2)
Institut de Ciencia de Materials de Barcelona,
Campus de la UAB,
08193 Bellaterra,
Catalunya,
Spain

J Santiso (X6.2)
Institut de Ciencia de Materials de Barcelona,
Campus de la UAB,
08193 Bellaterra,
Catalunya,
Spain

Mark D Savellano (X2.15)
Thomas J Watson Laboratory of Applied Physics,
California Institute of Technology,
Pasadena, CA 91125,
USA

M Sayer (X4.0)
Department of Physics,
Queen's University,
Kingston, Ontario,
Canada K7L 3N6

Jochen M Schneider (A5.1)
Thin Film Physics Division,
Department of Physics,
Linköping University,
S-581 83 Linköping,
Sweden

A Schrey (X1.0)
Institut für Werkstoffkunde,
Rhein Westfälische Technische Hochschule,
Aachen Templergraben 55 D-52056,
Aachen,
Germany

S Ismat Shah (A3.0, X1.8, X5.4)
E.I. DuPont de Nemours,
Experimental Station,
PO Box 80356,
Wilmington, DE 19898-0356,
USA

Tatsuo Shimizu (A1.5)
Department of Electrical and Computer Engineering,
Faculty of Technology, Kanazawa University,
Kanazawa 920,
Japan

Masanori Shinohara (X2.3)
NTT LSI Laboratories,
3-1 Morinosato Wakamiya,
Atsugi-shi, Kanagawa 243-01,
Japan

Hadas Shtrikman (X2.3)
Braun Center for Submicron Research,
Weizmann Institute for Science,
Rehovot 76100,
Israel

Siva Sivaram (X2.0)
Intel Corporation, SC11-104,
2200 Mission College Blvd,
Santa Clara, CA 95052-8119,
USA

Paul Snyder (D2.3)
Department of Electrical Engineering,
University of Nebraska,
Lincoln, NE 68588-0511,
USA

Zhizhong Song (X1.7)
Department of Physics,
Lanzhou University,
Lanzhou 730000,
People's Republic of China

Eberhard Spiller (F5)
60 Lakeside Road
Mt Kisco, NY 10549
USA

A J SpringThorpe (D2.0)
Bell-Northern Research,
PO Box 3511, Station C,
Ottawa, Ontario,
Canada K1Y 4H7

William D Sproul (A5.1)
Sputtered Films, Inc.,
320 Nopal Street,
Santa Barbara,
CA 93103,
USA

William D Sproul (X1.1, X1.2, X1.3, X1.4, X1.5, X1.6)
BIRL,
1801 Maple Avenue,
Evanston, IL 60201-3135,
USA

Christoph Steinbrüchel (C1.1, C1.2, C1.3)
Rensselaer Polytechnic Institute,
110 8th St.,
Materials Department, Rm 4225,
CII Building,
Troy, NY 12180,
USA

Samuel Strite (A2.4, X2.1)
IBM Research Division,
Zürich Research Laboratory,
8803 Rüschlikon,
Switzerland

Tuomo Suntola (B1.5)
Microchemistry Ltd,
PO Box 45,
Espoo, FIN-02152,
Finland

B Sverdlov (X2.10)
Coordinated Science Laboratory and Material Research Laboratory,
University of Illinois at Urbana-Champaign,
104 South Goodwin Avenue,
Urbana, IL 61801,
USA

J D Targove (X3.0, X3.5, X3.6)
Itek Optical Systems,
10 Maguire Road,
Lexington, MA 02173-3199,
USA

S Tatarenko (X2.4)
CEA-DRFMC/SP2M,
Laboratoire de Physique des Semiconducteurs,
17 avenue des Martyrs,
38054 Grenoble Cedex 9,
France

Patricia A Thiel (D1.1)
Department of Chemistry and Ames Laboratory,
Iowa State University,
Ames, IA 50011,
USA

R C Tucker Jr (A4.0, A4.2, A4.4, A4.5)
Praxair Surface Technologies, Inc.,
1500 Polco Street,
Indianapolis, IN 46224,
USA

R T Tung (X2.14)
AT&T Bell Labs,
600 Mountain Avenue,
Murray Hill, NJ 07974,
USA

Karl M Unruh (X2.9)
University of Delaware,
Department of Physics and Astronomy,
Newark, DE 19716,
USA

E M T Velu (X5.0)
Western Digital Corp.,
2109 Tasman Drive,
Santa Clara, CA 95054,
USA

Lili Vescan (B1.0, B1.4)
Institute for Thin Film and Ion Technology,
Research Center KFA,
Jülich D-5170,
Germany

Matthew M Waite (X1.8)
E.I. DuPont de Nemours,
Experimental Station,
PO Box 80356,
Wilmington, DE 19898-0356,
USA

C A Wang (X6.3)
Superconductive Materials Laboratory,
Acheson Hall,
State University of New York at Buffalo,
Buffalo, NY 14214,
USA

Jui H Wang (X6.3)
Superconductive Materials Laboratory,
Acheson Hall,
State University of New York at Buffalo,
Buffalo, NY 14214,
USA

I M Watson (X6.2)
Department of Chemistry,
Queen Mary and Westfield College,
Mile End Road,
London E1 4NS,
UK

I A Weerasekera (X2.9)
University of Delaware,
Department of Physics and Astronomy,
Newark, DE 19716,
USA

Timothy P Weihs (F7)
Department of Materials Science and Engineering
The Johns Hopkins University
3400 North Charles Street
Baltimore, MD 21218-2689
USA

Alex A Wernberg (X4.2)
Photoscience Research Division,
Imaging Research & Advance Development,
Eastman Kodak Company,
Rochester, NY 14650-2109,
USA

M R Wertheimer (E3.0)
Groupe des Couches Minces &
Department of Engineering Physics,
Ecole Polytechnique,
Box 6079, Station 'Centre-Ville',
Montreal, Quebec H3C 3A7,
Canada

William D Westwood (A5.0, A5.2)
7 Mohawk Crescent,
Nepean, ON,
K2H 7G7,
Canada

R Claude Woods (D2.6)
Department of Chemistry,
University of Wisconsin-Madison,
1101 University Avenue,
Madison, WI 53706,
USA

Ming-Show Wong (X1.1, X1.2, X1.3, X1.4, X1.5, X1.6)
BIRL,
1801 Maple Avenue,
Evanston, IL 60201-3135,
USA

X D Wu (X3.6)
Los Alamos National Laboratory,
Superconductivity Technology Center,
M/S K763,
Los Alamos, NM 87545,
USA

John Q Xiao (X5.0)
Department of Physics and Astronomy,
University of Delaware,
Newark, DE 19716,
USA

Masako Yudasaka (X7.2)
Yoshimura π-Electron Materials Project,
ERATO JRDC,
c/o Matsushita Research Institute of Tokyo,
3-10-1 Higashimita, Kawasaki 214,
Japan

C Zaldo (X3.4)
Instituto de Materiales de Madrid, CSIC,
Campus Universitario de Cantoblanco C-IV,
28049 Madrid,
Spain

Index

II–VI semiconductor superlattices F6.2:1
III–V compound D1.0:2, D1.0:8, D2.1:3, D2.4:5
III–V compound semiconductors B1.1:1, B1.1:12, D1.2:1, D2.0:1
III–V films A2.2:17
III–V materials B1.4:34
III–V semiconductor D2.1:1
III–V semiconductor superlattices F6.2:1
III–V substrates A2.2:17
IV–VI semiconductor superlattices F6.2:1

ALE of covalent (elemental) materials B1.5:2
ALE processing B1.5:3
ALE sequences B1.5:1
Abeyant sheath A3.1:17
Ablation lasers A1.5:2
Ablative photo-decomposition (APD) A1.5:2
Abnormal glow A3.0:2
Absorption D2.4:1
Absorption index F5:9
AC sputtering A5.2:2, A5.2:7
  reactive A5.2:2
Acceptor D2.4:5
  transitions D2.4:5
Acoustic impedance D4.0:2
Activation energy B1.5:4, B1.5:5, B1.5:7, F7:11
Activation energy barrier A2.3:2
Active storage environments E1.0:19
Adatom incorporation D1.0:5
Adatom migration D1.0:7
Additive C1.0:5
  ALE growth B1.5:6
  pattern transfer C1.0:3
Adhesion B1.2:5, B1.2:9, D0:2, E3.0:4
  enhancement E3.0:5, E3.0:10–12
  failure B1.2:9
  strength E3.1:19
Adiabatic temperature F7:9
Afterglow reactors B1.2:11
Aksenov device A1.4:12
$Al_xGa_{1-x}As$ D1.2:2, D1.2:3, D2.3:4
AlAs A2.0:17, D2.3:5
$AlAs_xSb_{1-x}$ D1.2:3
AlAs/GaAs superlattice D2.3:5
(Al, Ga)As A2.0:2
AlGaAs A2.0:17, A2.1:7, A2.3:5
AlInAs A2.0:17
AlSb A2.0:17
Alarm system A2.2:15
Alkaline cleaners E1.0:6
Alkane-based plasma chemistries C1.1:14
Allotropic forms A2.2:3
Alloy clustering D2.4:7
Alloy composition D0:6, D0:9, D1.2:1, D1.2:3, D2.0:3, D2.0:6, D2.3:4, D2.4:1
Alloy content D0:2
Alloy decomposition F1:7
Alloy reduced energy E2.2:9
Alternating current (AC) A5.2:1
Ambipolar diffusion A3.1:4, B1.2:13
Amperian sheet currents A3.2:5
Ancillary magnetic fields A3.2:21
Angle-resolved XPS (ARXPS) E3.1:11, E3.1:13, E3.1:18
Angular dependence of etch yields C1.2:3
Anisitropic etching C1.0:7, C1.1:3, C1.1:11
Anisotropic magnetoresistance (AMR) X5.0:14
Anisotropy C1.1:1
Antiphase domains A2.0:30
Antiphase boundaries D1.0:3
Arc evaporation A1.4:1
Arc evaporation sources A1.4:7
  ac A1.4:7
  dc A1.4:7
  filtered A1.4:7
  pulsed A1.4:7
  steered A1.4:7
Arc lamps B1.3:9
Argon (Ar) ion laser B1.3:11, D2.4:2
Arrhenius analysis F7:11
Arrival rate ratio A1.0:2
Arsine A2.2:1
As–As dimer D1.0:3
Aspect ratio C1.0:5, C1.0:6
Asperities D1.0:4
At rates B1.3:5
Atmospheric pressure glow discharge (APGD) E3.0:5, E3.0:30, E3.0:34
Atom diagnostics D0:2
Atomic absorption spectroscopy D0:3, D0:5, D0:6
Atomic arrangement of the surface D1.0:1
Atomic beams A2.0:1
Atomic bond energy F7:5
Atomic configurations D0:4

Atomic diffusion F7:8
Atomic intermixing F7:3
Atomic layer epitaxy A2.0:1
Atomic scattering factors F5:1, F5:2
Atomically abrupt interfaces A2.2:17
Auger electron spectroscopy (AES) D0:6
Automation A4.1:8
Available wavelength B1.3:11
Avalanche photodiodes F6.2:1
Average ion energy A1.4:2

Backscattering C1.0:8
Baffled furnace sources A1.1:4
Balanced and unbalanced magnetron fields A3.2:9
Bandgap D0:2, D0:3, D0:9, D2.4:1, D2.4:2
  energy D2.4:7
  engineering F6.0:11
Band-to-band transition D2.4:1, D2.4:5
Bar sources A1.1:4
Barriers to incorporation D1.0:7
Basic sections A1.2:1
Be dopant A2.3:6
Bias sputtering A3.0:1
Biasing schemes X5.0:13
Bidirectional reflectance distibution function D2.5:2
Binary compounds A2.0:14
Biocompatibility B1.2:5
Biot-Savart law A3.2:8
Bipolar transistor F6.0:13
Blocking capacitor A3.1:12
Boat sources A1.1:1
Boat/evaporation alloy A1.1:1
Bohm criterion A3.1:10
Bond coat A4.1:16
Bonding D0:7
Bound excitons D2.4:5
Boundary imperfections F5:6
Boundary layer B1.4:14
  models B1.4:13
Bragg condition F5:6
Bragg reflection D1.0:5
Bragg's law A2.0:11
Breadloafing B1.2:11
Bubbler A2.1:3
Bulk morphology D0:1
Busbar design A1.1:5

C^+ A3.4:17
Cable terminating lugs A1.1:5
Calculation methods B1.4:4
Calculation of thermodynamic equilibrium B1.4:2
Cantilevered beam D4.1:2
Capacitance bridge D4.1:4
Capacitive measurement D4.1:4
Cascade mixing E2.2:24
Carbide A3.4:17
Carbon contamination A2.2:6, A2.2:10
Carbon contamination of the layer B1.3:8
Carbon dioxide lasers B1.3:10
Carbon doping A2.2:1
  in GaAs A2.3:1
Carrier concentration D0:8
Carrier gas A2.1:3
Catalysis B1.0:10
Catalytic cracker A2.1:3
Cathode dark space A3.0:2, A3.1:2
Cathode erosion rate A1.4:2
Cathodoluminescence D2.4:2
Cauchy oscillator model D2.3:2
(Cd, Hg)Te A2.0:2
CdHgTe A2.0:21
CdTe A2.0:21
Characteristic temperature A3.1:9
Charge exchange A3.4:22
Charge form A1.2:7
Charge transport B1.2:11
Chemical activation B1.2:1
Chemical beam epitaxy (CBE) D2.0:1
Chemical beam etching A2.3:9
Chemical bonding D0:2, D0:6, D0:9
Chemical composition D0:6
Chemical kinetics B1.4:15
Chemical memory B1.2:16
Chemical ordering D0:2
Chemical processes used in CVD B1.0:5
Chemical profile F7:4
Chemical transport B1.0:9
Chemical vapour deposition (CVD) B1.2:7, D0:2, D0:4, D0:8
  equipment B1.4:21
  sources B1.4:8
  techniques B1.0:1
Chemically assisted ion beam etching C1.2:1
Chemically enhanced physical sputtering C1.2:4
Child's law A3.1:5
Chlorofluorocarbon (CFC) solvents E1.0:5
Cl- and Br-based plasmas C1.1:5, C1.1:12
Classification of ALE reactors B1.5:9
Cleaning
  abrasive E1.0:2
  chamber B1.2:15
  external E1.0:5
  gross E1.0:2
  handling and storage aspect of E1.0:17
  hydrogen plasma E1.0:10
  in situ E1.0:1, E1.0:21
  megasonic E1.0:13
  oxidative E1.0:8
  reactive gas E1.0:8
  reactive plasma E1.0:9, E1.0:21
  solvent E1.0:4
  specific E1.0:4

sputter E1.0:21
substrate E1.0:1
ultrasonic E1.0:12
wet reaction E1.0:7
Cleaning tests E1.0:19
Clearance control A4.1:15
Cluster effects E3.1:14
Cluster tools B1.4:25
Clustering D0:4
$CoSi_2$ A2.0:25
Cohesive energies E2.2:17
Coil sources A1.1:1
Cold spray A4.1:3
Cold-wall reactor B1.4:22, B1.4:23
Collection probability A3.4:4
Collision cascade A3.4:2, E2.2:24
Columnar growth A1.4:6
Columnar structure A3.4:12
Colutron source A2.4:3
Combined ion beam and molecular beam epitaxy CIBMBE A3.4:20
Combined reactions B1.0:11
Combustion spraying A4.1:1
Commercial profilometer D4.1:5
Compensation point A3.3:6
Complex reflectance D2.1:1
Complex refractive index D2.3:1
Composition A2.2:17, D0:2, D0:3, D0:7, D1.2:2, D2.0:7, D2.3:3, D2.3:5
profile D1.2:2, F7:10
Composition-depth profiles A2.2:11
Compositional profile D0:3, D1.2:3
Compound formation reactions F7:1
Compound reactants B1.5:6, B1.5:7
Compound semiconductors C1.1:13, D0:6
Compounds A1.0:11
Compression ratio A2.3:13
Conditioning A1.0:11
of the evaporant A1.2:6
Conduction band D2.4:1
Conduction heat transfer E2.2:26
Conformal films B1.2:13
Conical magnetrons A3.2:1
Contact printing C1.0:1
Contamination A2.2:6, A2.2:10
Contrast D2.2:6
Control the stress D4.1:7
Convective systems B1.2:3
Cord flame spray A4.1:7
Corona treatment E3.0:5, E3.0:30
Corrosion current density A3.4:12
Corrosion resistance A3.4:10, A4.1:15
Corrosion resistant layers A1.4:7
cos θ distribution A1.5:3
Cosine distribution A3.0:12
Cost A2.2:2
Coulomb explosion A1.5:3
Counter electrode A3.1:12
Cracker cells A2.0:5, A2.0:7
Cracker evaporation sources A2.0:2
Cracking efficiency A2.2:10
Cracking patterns A2.2:3
Cracking temperatures A2.3:1
Criteria for choosing reactants B1.5:13
Critical angle F5:4
Critical curvature D4.1:2
Critical point structure D2.3:3, D2.3:4
Crook's dark space A3.1:2
Cross contamination A2.2:18, B1.2:7
Cross sections B1.2:2
Cross-sectional transmission electron microscopy D0:4
Crosstalk C1.0:8
Crucible heaters A1.1:4
Cryo-shielding A2.3:11
Cryostat D2.4:2, D2.4:3
Crystal monitor D0:4, D0:5, D0:6, D2.3:3
Crystal orientation D0:1, D0:3
Crystal perfection D1.0:5
Crystal structure D0:1, D0:3
Crystallographic orientation D1.0:3
Crystallographic orientation D4.0:2
Cu oxide A3.4:21
Current-voltage (*I–V*) characteristics D3.0:1
Cycling A2.2:14
Cylindrical magnetrons A3.2:1
Cylindrical-hollow magnetron A3.0:2

Damage C1.3:1, D1.2:3
Damage energy E2.2:22
Damping (internal friction) F2:7
Dark conductivity D0:2, D0:8
Data acquisition for LEED experiments D1.2:3
Debye length D3.0:2
Debye–Waller factor F5:5
Deceleration A3.4:9
Decorative coatings A1.4:7
Deep UV lithography C1.0:1
Defect centres B1.2:11
Defect concentration D0:2, D0:8
Defect structures D1.0:1
Defects D0:4
Deflection of thin plates D4.1:2
Density D0:1, D0:4, D0:6, D2.3:3
Dependence on gas pressure D4.0:5
Deposition D2.0:2
efficiency B1.4:5
of alloys A1.0:3
rate A1.3:6, B1.2:7, D0:4, D0:5, D2.3:3
Depth of modification E3.1:11
Design of x-ray mirrors F5:6
Desorbed flux D1.2:1

Desorption mass spectrometry (DMS) D1.2:1
Detergent (soap) cleaners E1.0:6
Devices A2.2:3
Diamond B1.2:5, D0:6
Diamond-like carbon (DLC) A1.4:7
  film A3.4:2
Dielectric function D2.3:2, D2.3:3, D2.3:4
Dielectrics D2.3:3
Differential charging E3.1:14
Diffraction limited spot size B1.3:12
Diffraction pattern D0:8, D1.0:1
Diffusion B1.2:3
  atomic F7:8
  barriers B1.2:5
  thermal F7:8
Dimer bonds D1.0:2
Dimerization A2.2:10
  efficiency A2.2:11
Diode reactors B1.2:8
Diode systems B1.2:11
Direct bandgap D2.4:1
  materials D2.4:2
Direct ion beam deposition DIBD A3.4:1
Direct-write methods C1.0:6
Direction of temperature change D2.2:5
Directional probes D3.0:7
Directionality of etching processes C1.0:7
Discharge centreline A3.1:4
Dislocation image force F2:10
Dislocations F2:2
Disorder D0:4, D2.3:3
Displaced atoms E2.2:22
Displacement energy E2.2:21
Displays B1.2:5
Disproportionation B1.0:9
Dissociative ionization B1.2:2
Divergence A3.4:9
Domain structure D1.0:3
Donor D2.4:5
  transitions D2.4:5
Dopant concentration D0:2, D0:8
Dopant profiles A2.0:20
Doping A5.0:1, A5.0:4
Dose of reactant needed B1.5:15
Downstream reactors B1.2:11
Driving force F7:6
Dry etching C1.0:7
Drying E1.0:16
DSC studies F7:11
Dual magnetron systems A3.2:21
Duoplasmatron A3.3:3
Dynamic ion beam mixing A3.4:1
Dynamics of thin film growth D1.0:1

$E \times B$ Wien filter A3.4:22
Eddy current loss X5.0:9
Effect of source to substrate distance A1.0:5
Effect of temperature B1.5:7
Effective discharge resistance A3.1:15
Effective electron emission coefficient A3.1:3
Effective index F5:4
Effective life D4.0:3
Effects of thermal expansion A1.1:1
Efficiency of cracking A2.2:10, A2.2:11
Effusion oven A2.2:1
Effusion rate A2.0:5
Ejected clusters A1.5:2
Elastic behaviour F2:3
Elastic collisions E2.2:5
Elastic scattering B1.2:2
Electric stopping E2.2:4, E2.2:5
Electrical diagnostics D0:2
Electrical properties D0:8
Electrochemical deposition D0:2
Electroluminescence D2.4:2
Electromagnetic isotope separator A3.4:2
Electromechanical D4.1:5
  diagnostics D0:2
  probe D0:4
Electron beam D1.0:1, D1.0:3, D1.0:4
Electron beam heated evaporation sources A1.2:3
Electron beam power supply A1.2:3
Electron channels A3.2:17
Electron confinement D2.4:5
Electron cyclotron resonance (ECR) etching C1.1:3, C1.1:17
  in O-based plasmas C1.1:18
  of compound semiconductors C1.1:19
  of metals C1.1:18
Electron cyclotron resonance (ECR) ionization A3.0:1
Electron cyclotron resonance (ECR) plasma sources A2.1:16, B1.2:12
Electron cyclotron resonance (ECR) source B1.2:12, C1.1:15
Electron density A1.4:3
Electron diagnostics D0:2, D0:4, D0:8
Electron effective mass D2.4:4
Electron emitter A3.4:17
Electron energy distribution B1.2:2
  function (EEDF) E3.0:6
Electron gun A1.2:4
Electron gyroradius D3.0:2
Electron impact ionization A3.3:1
Electron microscopy D0:8
Electron mobility B1.2:5
Electron plasma density D3.1:1
Electron plasma frequency D3.1:1
Electron temperature A3.2:3, B1.2:2, D3.0:1
Electron traps A3.2:17
Electron velocity D3.0:1
Electron wavelength D1.2:1

Electronic collisions E2.2:4
Electronic diagnostics D0:2
Electronic transitions D0:6
Electronic traps B1.2:11
Elemental reactants B1.5:4
Eliminating macroparticles A1.4:4
Ellipsometer D2.3:4, D2.3:6
Ellipsometry D0:5, D0:6, D0:7, D2.1:1, D2.1:2, D2.1:4
Elliptical distortion D4.1:2
Emission intensity B1.3:9
Emission interferometry D0:5
Emissions from the source A1.2:2
Emissive probes D3.0:5
Emissivity D0:7, D2.0:4
End-point detection D0:1, D0:5, D2.2:1
Energetic ions E2.2:1
 range distribution of E2.2:1
Energy barrier D1.0:7
 to hopping D1.0:7
 to incorporation D1.0:8
 to surface migration D1.0:5
Energy distribution functions D3.0:1
Energy input required for vaporization A1.0:2
Energy levels D2.4:1
Energy-loss processes E2.2:3
Energy spread A3.4:4
Energy transfer factor E2.2:15
Enhanced diffusion rate B1.3:2
Enhanced discharges C1.1:14
Enhancement mechanisms C1.1:2
Enthalpy of formation F7:5
Enthalpy of melting F7:7
Epitaxial film growth A3.0:13
Epitaxial growth A3.4:15, D2.4:1, D2.4:2, D2.4:5
Epitaxial processes D2.1:4
Epitaxial silicon B1.4:31
Epitaxial temperature A3.0:13
Epitaxy A2.0:1
Equilibrium constant B1.4:2
Equilibrium partial pressure B1.4:4
Equilibrium-controlled process B1.0:4
Equipment for ion plating A1.3:1
Equipment needed for end-point interferometry D2.2:2
Erosion track A3.2:4
Etch anisotropy C1.1:6
Etch mechanism C1.1:1
 for polymers C1.1:11
Etch rate C1.1:1, C1.1:11, C1.1:14, C1.1:17
 of SiO_2 C1.1:9
 of Si C1.1:9
Etch selectives C1.2:1
Etch selectivity C1.1:11
 of Si over SiO_2 C1.1:8
Etch yield C1.1:10, C1.1:17, C1.2:3, C1.2:4
Etching gas A2.3:8
Etching of SiO_2 C1.1:10
Etching of Ge and Si–Ge alloys C1.1:8
Etching of Si in Br-containing plasmas C1.1:8
Etching of Si in F-containing plasmas C1.1:6
Etching of carbonaceous materials C1.2:3
Etching rate D2.2:1
Etching with neutral beams C1.2:6
Evaporable form A1.0:12
Evaporant alloy composition A1.0:4
Evaporant distribution A1.0:4
Evaporation B1.2:6
 of alloys A1.0:13
 rate A1.2:2
Ewald sphere D1.0:1
Exchange reaction B1.5:4, B1.5:6
Excimer lasers B1.3:11
Excitation
 rotational B1.2:2
 vibrational B1.2:2
Exciton D2.4:1, D2.4:4, D2.4:5
 binding energies D2.4:5
 complexes D2.4:5
 recombination D2.4:1
 reduced mass D2.4:4
Explosive ejection A1.5:2
Explosive emission A1.4:5
Extant sheath A3.1:17
External hardware D4.0:2
Externally controlled shutters D4.0:3
Extinction coefficient D0:2, D0:6, D0:8
Extraction voltage A3.3:3

F- and Cl-based plasmas C1.1:13
F-based plasmas C1.1:5
Faraday collector D1.0:1
Faraday dark space A3.1:2
Faraday rotation A1.5:8
$^{56}Fe^+$ A3.4:11
Fe_2O_3 A3.4:13
Fe(110) surface A3.4:13
Fe film A3.4:10
Feasibility of a reaction B1.4:3, B1.4:8
Fermi level D0:2, D0:8
Ferrite heads X5.0:11
Field ionization A3.3:1
Film composition A2.2:1
Film delamination A3.0:2
Film density and adhesion D4.0:5
Film growth in GSMBE A2.2:1
Film morphology D1.0:5
Film thickness D0:1
Film uniformity A1.0:11
Filtered sources A1.4:11
Flame spray A4.1:1
Flash evaporation A1.0:4

Flat panel display A3.4:17
Flat substrate holder A1.0:7
Floating potential A3.1:11
Flow regimes B1.4:11
Flow transients A2.2:8
Flux from real electron beam heated sources A1.0:7
Flux transient A2.2:13
Focused ion beam FIB A3.4:22, C1.3:1
Focused laser irradiation A1.5:1
Folded box boats A1.1:4
Formation reactions F7:1, F7:2
  exothermic F7:2
Forward recoil sputtering A1.4:6
Fracture F2:11
Frank–van der Merwe growth A2.0:25
Free exciton D2.4:1, D2.4:4
  luminescence D2.4:4
Free space velocity D3.1:1
Freeman type ion source A3.4:7
Frequency B1.2:10
  shift D4.0:2
Fresnel equation F5:3, F5:4
Fresnel reflection coefficients D2.3:1
Full dynamical analysis D1.2:2

Ga beam cleaning A2.2:17
$Ga_{1-x}Al_xAs$ D2.0:1, D2.0:2, D2.0:3, D2.0:5
Ga-alkyls A2.3:3
Ga-stabilized surface A2.3:2
GaAs (001) 2 × 4 reconstructed surface D1.0:3, D1.0:6, D1.0:7
GaAs (001) surface D1.0:2, D1.0:7
GaAs (100) D1.0:11
GaAs A2.0:2, A2.3:1, D1.0:7, D1.0:8, D1.2:3, D2.1:3, D2.3:5, D2.4:4, D2.4:5, D2.4:6
$GaAs_xSb_{1-x}$ D1.2:3
GaAs–AlAs superlattice D2.3:4
GaAs/AlAs D2.0:1
  alloy systems D2.0:7
  interface D1.2:3
GaAs/AlGaAs interface D1.2:3
GaAs/InGaAs interface D1.2:3
(Ga, In)As A2.0:2
GaInAs A2.0:17, D2.1:4
GaInAsP A2.3:1
GaInAsSb A2.0:17
GaInP A2.0:17
GaSb A2.0:17
Gas cracker A2.2:18
Gas cracking oven A2.2:10
Gas delivery system A2.2:18
Gas dynamic theory A1.4:2
Gas filter A2.2:13
Gas flow C1.0:10
Gas injector cells A2.1:1
Gas injector nozzle A2.3:1
Gas injectors A2.3:11
Gas lines A2.2:8
Gas phase nucleation B1.2:8
Gas residency B1.2:14
Gas scrubber A2.2:3, A2.2:16
Gas storage cabinets A2.2:15
Gas-phase dissociation of silane B1.4:17
Gas-source MBE A2.0:1
Gaseous feed stocks A2.2:1
Gauss' law A3.1:5
Ge detectors D2.4:2
Ge film A3.4:15
Gettering A5.0:5
Giant magnetoresistance (GMR) F4:1, X5.0:8
Gibbs free energy of the reaction B1.4:2
Glow discharge A3.1:1, B1.2:1, C1.0:9, C1.0:10
  ion source A3.3:3
  plasma B1.2:2
  sputtering A3.0:1
  treatment E3.1:1
Graded composition D0:6
Grain size D0:1, D0:3, D0:6
Growth dynamics D1.0:1, D1.0:8
Growth kinetics B1.4:12
Growth mechnisms B1.0:3, B1.4:11, D2.1:1
Growth mode D1.0:8
Growth mode transition D1.0:5
Growth procedure A2.2:17
Growth processes F1:1
Growth rate A2.2:1, A2.2:13, B1.3:5, B1.3:15, B1.4:13, B1.4:18, D0:3, D0:8, D2.0:2, D2.0:5, D2.0:7, D2.3:5
  temperature regions B1.0:3

Halides B1.4:8
Hall acceleration A1.4:11
Hall–Petch effect F2:9
Hard and protective films B1.4:37
Hardness D0:2, F2:7
HBT F6.0:13
Hearth assembly A1.3:5
Heat capacity F7:7
Heat of crystallization F7:4
Heat of formation ΔH_f F7:9
Heat of reaction ΔH_{rx} F7:5
Heat sink A3.4:17
Helicon source C1.1:15
Helium–neon (HeNe) laser D2.4:2
Hemispheric magnetron A3.2:1
Heteroepitaxial D2.0:1
  deposition D2.0:2
Heteroepitaxy A2.0:25
Heterojunction F6.0:12
Heterojunction formation D2.1:3
Heterostructure deposition D2.0:7
Heterostructures A2.0:1

$Hg_xCd_{1-x}Te$ D2.3:4
High energy neutral A3.4:22
High conductance crackers A2.2:11
High purity A3.4:1
High resolution electron energy loss spectroscopy (HREELS) E3.1:3
High temperature superconducting material A3.4:21
Hittorf dark space A3.1:2
Hole effective mass D2.4:4
Hollow cathode discharge A3.1:11
Homoepitaxial growth A3.4:15
Homoepitaxy A2.0:25
Homogeneous and heterogeneous reactions B1.1:7, B1.4:20
Horizontal reaction cells B1.3:14
Horizontal reactor B1.1:8
Hot-wall reactors B1.4:21
Hydride cracker A2.2:10
Hydride cracking cells A2.3:12
Hydride delivery systems A2.2:7
Hydride gas detectors A2.2:15
Hydrides A2.3:1, B1.4:8
Hydrogenated amorphous silicon B1.2:5
Hydrolysis B1.0:8
  of imines E3.1:13
Hysteresis A5.0:5

i-carbon film A3.4:17
I–V curves D1.2:2, D1.2:5
IBACE of SiO_2 C1.2:4
Ignition of self-propagating reactions F7:4
Impingement rate ratio A1.0:1
Implantation distribution E2.2:13, E2.2:16
  low dose regime E2.2:13
  high dose regime E2.2:16
Impurity A3.4:12
  content D0:2
  gettering A2.1:7
  levels D2.4:5
  transition D2.4:5
In situ diagnostics B1.3:15
In situ etching A2.3:8
In situ measurements D0:1, D0:2, D0:3
In situ monitor D2.5:4, F5:10
In situ observation B1.5:10
In situ reflectivity F5:13
$In_xAl_{1-x}As$ D1.2:3
$In_xGa_{1-x}As$ D1.2:3, D2.3:4
InGaAs A2.3:3
InGaAsP A2.3:7
InP A2.0:17, C1.3:1, D2.1:3
InP/GaInAs D2.1:4
InP/GaInAs interface D2.1:3
InSb A2.0:17
Incorporation of the sputtering gas A3.0:2
Incorporation rate A2.0:19, D0:7, D1.0:7
Index of refraction D0:2, D0:5, D0:8
Indirect bandgap D2.4:1
  materials D2.4:2, D2.4:5
Indirect transitions D2.4:5
Indium-free mounting A2.0:14
Indium tin oxide X3.6:2
Inductive heads X5.0:8
Inductively coupled sources C1.1:16
Inelastic collisions E2.2:5
Inelastic scattering D1.0:4
Influence of temperature D4.1:5
Information available from LEED D1.2:2
Infrared (IR) laser interferometric thermometry D2.2:4
Insulating targets A5.0:1, A5.2:1, A5.2:3, A5.2:7
Interdiffusion F2:10
Interdiffusion coefficient F7:9
Inter-electrode distance B1.2:14
Interfaces A3.4:21, B1.2:8
Interfacial charge B1.2:11
Interfacial chemistry E3.1:14
Interfacial interdiffusion D0:2
Interference effects D2.0:1, D2.0:4
Interference fringes D2.0:3
Interference reflectometry D2.0:5
Interference spectra D2.0:4
Interferograms D2.0:7
Interferometer D3.1:1
Interferometric thermometry D2.2:5
Interferometry D0:5, D0:6, D0:7, D0:8, D0:9
Interferrometric techniques D4.1:3
Interlayer coupling F4:5
Intermetal dielectrics B1.2:13
Intrinsic (growth) stress F2:2
Inverted magnetron A3.0:2
Ion assisted deposition A1.0:3
Ion beam C1.0:9, C1.0:10
  assisted deposition IBAD A3.4:1
  deposition IBD A3.4:1
    combined A3.4:19
    maskless A3.4:8
    negative A3.4:8
  mixing E2.2:22
  processing E2.2:1
  source E3.0:19
  sputter deposition IBSD A3.4:1
  sputtering (IBS) A3.0:1, A3.3:1
Ion bombardment A3.4:1, B1.2:4, B1.2:9, C1.1:1
Ion damage E2.2:21
Ion diagnostics D0:2
Ion doping A3.4:20
Ion emission A1.4:2
Ion energy distribution function (IEDF) E3.0:6
Ion fractions A1.4:6
Ion flux A3.4:4, B1.2:13
Ion implantation A3.4:15

Ion milling C1.0:10
Ion plating A1.0:3
Ion saturation current D3.0:5
Ion scrubbing E1.0:21
Ion stopping E2.2:3
Ion vapour deposition source A1.3:2
Ion–solid interaction A3.4:25
Ion–surface interactions A3.0:14
Ion-assisted IBS A3.0:1
Ion-beam-assisted chemical etching C1.2:1, C1.2:4
Ion-beam-induced damage C1.2:1
Ion-peening process A3.0:2
Ionization cross section A3.0:3
Ionized cluster beam deposition ICB A3.4:1
Ionized fraction A1.4:11
Ions for ion-beam-assisted deposition A3.3:1
Islands F1:5
Isotope A3.4:1
Isotopically enriched A3.4:12
Isotropic etching C1.0:7
ITO X3.6:2

Kaufman ion beam source A2.4:1, A3.0:6, C1.2:1
Kelvin probe D0:2, D0:8
Kikuchi lines D1.0:5
Kikuchi patterns D1.0:5
Kikuchi-line features A2.0:11
Kinetic energy of ions A3.1:16
Kinetic-limited regime B1.0:4
Knudsen cell A2.0:3, A2.2:1
Knudsen number B1.4:11

LCVD reaction chambers B1.3:12
LEED equipment and operations D1.2:2
LEED operating tips D1.2:5
Lambert's cosine law A3.2:14
Langmuir evaporation A2.0:14
Langmuir probe A3.1:9, D0:2
Larmor equation A3.2:3
Larmor potential A3.2:16
Larmor radii A1.4:11
Laser evaporation A1.5:6
Laser interferometry D2.0:4, D2.2:3
Laser precimeter D0:8, D4.1:4
Laser reflectometry D2.0:4
Laser safety B1.3:12
Laser sputtering A1.5:6
Laser writing B1.3:6
Laser MBE A1.5:10
Lattice disorder E2.2:1
Lattice mismatch D1.0:8
Lattice-matched structure D2.1:4
Lattice-mismatched semiconductor D2.1:3
Lattice-mismatched structure D2.1:4
Law of mass action B1.4:3
Layer-by-layer growth A2.0:25, D0:4, D2.0:1, D2.3:3
Lift-off process C1.0:4
Ligand desorption B1.3:8
Ligands B1.5:7
Light scattering D0:5, D0:8, D2.5:1
Line average density D3.1:2
Linear photodiode array D0:6
Line-of-sight B1.2:6
Liquid phase epitaxy A2.0:2
Load-locked system B1.4:26
Loading effect C1.1:6
Local oscillator beam D3.1:1
Locus of failure E3.1:18
Long-working-distance microscope D4.1:3
Lorentz force A3.2:14
Low energy knock-on process A3.0:10
Low thermal budget B1.4:31
Low-angle grain boundaries A2.0:30
Low-energy electron diffraction (LEED) D0:4
Low-pressure cells B1.3:12
Low-pressure cracker A2.1:3, A2.2:10
Low-pressure MOCVD B1.1:2
Low-strain window D2.1:3
Luminous cathode spots A1.4:2

Machine element repair A4.1:13
Macroelectronics B1.2:5
Macromolecules B1.2:8
Macroparticle emission A1.4:4
Macroparticle erosion rate A1.4:4
Macroparticle filters A1.4:6
Macroparticle reduction A1.4:11
Magnetic lens A1.2:5
Magnetic materials C1.3:1
Magnetic plasma duct A1.4:13
Magnetic superlattices F4:1
Magneto-optical Kerr effect D2.1:1
Magneto-photoluminescence D2.4:1
Magnetoresistance X5.0:8
Magnetoresistive (MR) read heads X5.0:12
Magnetron ion etch (MIE) reactor C1.1:15
Magnetron ion etching C1.1:3, C1.1:16
Magnetron ion gauges A3.2:1
Magnetron sputtering A3.0:1, A5.0:1, A5.1:1
  pulsed A5.1:1, A5.1:2
  reactive A5.1:1, A5.1:5
  sources F5:10
Maintenance A2.2:14
Maskless etching or deposition C1.0:6
Mass diffusivity F7:9
Mass flow control A2.2:7
Mass separation A3.4:1
Mass spectrometer D1.2:1
Mass transport B1.4:13, B1.4:20
  regime B1.0:4

Mass-related sputtering effect A3.3:5
Matching of the power supply to the load A1.1:6
Material purity D2.4:4
Materials efficiencies B1.2:7
Matrix methods F5:6
Matrix notation D1.2:2
Mean free path A1.0:1, A2.0:1, A2.3:10, A3.0:2, D3.0:2
Measurable phase shift D3.1:1
Mechanical diagnostics D0:2
Membrane substrate D4.1:6
Metal alkoxides X3.5:2
Metal alkyl B1.1:3, B1.1:5
Metal carbonyls B1.3:3
Metal film A3.4:10
Metal-in-gap (MIG) X5.0:11
Metal oxides X3.5:2
Metallic mode A3.0:14
Metallic silicides B1.4:36
Metallization E3.1:16
Metalorganic B1.1:1
Metalorganic chemical vapour deposition (MOCVD) B1.1:1, B1.1:2, D2.1:1, D2.1:3, D2.3:3, F6.2:4, X3.5:2
Metals D2.3:3
Metastable alloy films A3.0:14
Metastable material A3.4:2
Microcrystalline A3.4:13
Microdroplets A1.4:2, A1.4:4
Microelectronics B1.2:5
Microroughness C1.0:9
Microstructure D0:2
Microtopography of the surface D2.5:1
Microwave absorption D0:2, D0:8
Microwave frequency D3.1:1
Microwave ion source A3.4:8
Migration A3.4:2
  of adatoms D1.0:8
Migration enhanced epitaxy A2.0:1
Minimum focus diameter B1.3:11
MIS capacitor A3.4:19
Misfit dislocations A2.0:29
Misfit strain F1:1
Mode of film growth D0:1
Molecular beam epitaxy (MBE) D0:2, D0:4, D0:8, D1.0:1, D1.2:1, D1.2:2, D2.0:1, D2.3:3, D2.3:4, D2.3:6, D2.4:2, D2.4:5, D2.4:7, F6.2:4
Molecular ion A3.4:9
Momentum transfer D4.1:5, F5:4
Monte Carlo simulations D1.0:1, D1.0:5, D1.0:7
Morphology A3.4:12
Moving substrate reactor B1.5:12
Multi-quantum well layers F6.2:1
Multi-shot reactor B1.5:9
Multihearth source A1.2:5
Multilayer A3.4:21
  C/W A3.4:21
  materials F7:1
  sample D2.3:2
  mirrors for x-rays F5:1
Multiple-detector array D2.3:5
Multiple-layered thin film D0:1
Mumetal X5.0:10

Nanoindentation F2:7
Narrow band oscillator D3.1:1
Nd-YAG lasers B1.3:11
Near bandgap transitions D2.4:1
Negative glow A3.0:3, A3.1:2
Negative ions A5.0:1, D3.0:7
Negative self-bias potential A3.0:5
Neutral dissociation B1.2:2
Newton's rings D4.1:3
$NiSi_2$ A2.0:25
Nitridation B1.0:8
Non-congruent evaporation A2.0:14
Non-polar solvents E1.0:4
Normal-flow reactors B1.4:22
Normal-incidence reflectance D2.1:1
Nuclear collisions E2.2:3
Nuclear stopping E2.2:4, E2.2:5
Nucleation D0:2
  and growth E3.1:16
  sites E3.1:15

O_2 or H_2 additions C1.1:6
Off centre sources A1.0:11
Operating pressure C1.0:9
Optical anisotropy D2.1:4, D2.3:3
Optical applications B1.2:5
Optical constants in x-ray region F5:1, F5:2
Optical diagnostics D0:2
Optical eigenaxes D2.1:1
Optical emission control A5.1:6
Optical emission spectroscopy D0:5, D0:8
Optical interference D2.2:2
Optical interferometry D0:3
Optical probes D0:5
Optical properties D0:7
Optical pyrometer D2.0:1
Optical receivers F6.0:1
Optical window D2.1:2
Optics of layered structures F5:5
Optoelectronics device A3.4:17
Ordering F1:2
Organo-metallic VPE A2.0:2, B1.1:1
Outdiffusion E1.0:17
Outgassing E1.0:17
Oxidation of Si in the presence of water B1.4:24
Oxidation reactions B1.0:7

Oxide desorption D1.2:3
Oxygen contamination A2.2:13

P_2 molecular beam A2.2:5
Parallel illuminated geometry B1.3:12
Parallel plate discharge D3.1:2
Parallel-flow reactors B1.4:22
Partial pressure A1.0:1
Particle energy distribution A2.4:22
Particle size B1.2:8
Particulate contamination B1.2:8, E1.0:14
  process-induced B1.2:8
Particulate formation B1.2:7
Particulate inclusions D0:2
Particulates D0:8
Paschen's law A3.0:4
Passive arc confinement A1.4:8
Passive storage environments E1.0:18
Pattern generation C1.0:1
Pattern replication C1.0:1
Pattern transfer C1.0:2
Patterning of Cu C1.1:13
Patterning of hydrocarbon polymers C1.1:11
Patterning of metals C1.1:12
Patterning of refractory metals C1.1:13
PbS detectors D2.4:2
Peach–Koehler force F2:10
Peak absorption wavelength B1.3:3
Peak temperature B1.3:2
  rise B1.3:1
Pearson transformer D3.0:9
PECVD films B1.2:4
Penning conditions A3.2:2
Penning discharge A3.1:1
Penning ion source A3.3:2
Periodic step array D1.0:4
Periscopic viewport D2.0:4
Permalloy X5.0:10
Phase D0:3, D0:6
  detecting circuit (PLL) D3.1:4
  diagram B1.4:5, B1.4:6, B1.4:8
  transformation F7:7, F7:11
Phase diagrams B1.4:6
Phase-sensitive detection D4.1:5
Phased locked epitaxy A2.0:1
Phonon D2.4:1, D2.4:5
  energy D2.4:7
Phosphine A2.2:1
Phosphorus contamination A2.2:5
Phosphorus fire A2.2:16
Photo-excitation of electrons D2.4:1
Photochemical quantum efficiency B1.3:3
Photoconductivity B1.2:11, D0:2, D0:8
Photodecomposition A1.5:5
Photodetector D2.4:2
Photodissociation B1.3:3
Photoelastic modulator D2.1:2
Photoinduced charge transfer B1.3:5
Photoluminescence D0:6
Photolysis B1.0:10, B1.3:2, B1.3:11
Photomultiplier tubes D2.4:2
Photon emission A1.4:2
Photosensitization B1.3:4
Photovoltaics B1.2:5
Physical vapour deposition (PVD) D0:4, D0:8
Pinhole D0:2, D0:8
'Pirhana solution' E1.0:7
Planar magnetrons A3.2:1
Planar probes D3.0:2
Planarization layers B1.2:13
Plasma electron density n_e D3.0:1
Plasma-enhanced chemical vapour deposition (PECVD) B1.2:1
Plasma equation A3.1:9
Plasma etching D0:5, E3.0:2, E3.0:3
Plasma jet A2.4:1
Plasma potential V_p D3.0:1
Plasma potentials A3.2:21, B1.2:15
Plasma processes D0:2
Plasma sources
  downstream E3.0:18
  dual mode microwave/radiofrequency E3.0:17
  electron cyclotron resonance E3.0:17
  microwave E3.0:15
  radiofrequency E3.0:14
Plasma surface modification E3.0:3
  ablation E3.0:8
  chemical modification E3.0:3, E3.0:10–12
  cleaning E3.0:7
  crosslinking E3.0:8
Plasma treatment of
  flexible substrates E3.0:21
  large objects E3.0:23
  small parts E3.0:27
Plasma-assisted MBE A2.0:2
Plasma-enhanced CVD D0:3, E3.0:2, E3.0:3
Plasmas D0:8
Plasma–surface interactions E1.0:10
Plastic behaviour F2:7
Plate theory D4.1:2
Plume A1.5:1
Poisoned cathode A3.0:14
Poisoning of the cathode A3.2:24
Poisson's ratio D4.1:2
Polar solvents E1.0:4
Polarization D2.3:1
Polarization modulation ellipsometry (PME) D2.3:5
Polycrystalline Si B1.4:32
Polymer C1.1:11
Positively charged ions A1.4:2
Post-ablation ionization (PAI) A1.5:3

'Potential hump' theory A1.4:2
Powder B1.2:8
  flame spraying A4.1:1
Power B1.2:10
Power supply A1.1:6
  for ion vapour deposition A1.3:2
Power spectral density F5:10, F5:13
Power spectrum of thin film roughness F5:14
Power welding A4.1:4
Precursor state B1.5:4
Precursors B1.2:6, B1.3:6
Pressure control A2.2:7
Primary ion beam deposition PIBD A3.4:1
Primary knock-on A3.0:10
  atoms E2.2:21
Probing beam D3.1:1
Process model D0:1, D0:3, D0:8
Process pressure B1.2:10
Processing equipment B1.5:8
Profilometer D2.0:5
Projected range E2.2:3, E2.2:9
  transverse E2.2:6
Projection imaging B1.3:6
Proximity discharge reactors B1.2:11
Proximity printing C1.0:1
Pseudodielectric function D2.3:2, D2.3:4
Pulmonary edema A1.5:10
Pulsed biasing D3.0:2
Pulsed laser deposition (PLD) A1.5:6, X3.6:2
Pulsed plasmas B1.2:14
Pumping system B1.4:28
Purification A3.4:12
Purity A2.2:13
Pyrolysis B1.0:5, B1.3:1, B1.3:2
  temperature B1.3:2, B1.3:8
Pyrometer D2.0:3, D2.0:4, D2.4:1
Pyrometry D0:7, D0:9, D2.0:4

Quadrupole mass spectrometer (QMS) D0:7, D1.2:1
Quantum dot F1:2
Quantum well D2.4:5, D2.4:7
  heterostructures A2.2:17
  structures B1.4:33
Quarter wave stack F5:7
Quartz crystal D0:4
Quartz microbalance D0:4, D0:8
Quartz oscillator D0:2

RA spectrometer D2.1:2
RDS D2.1:4
RIBE of Si C1.2:3
RIBE of Si, SiO_2 and Si_3N_4 C1.2:2
RIE of Si in Cl-based plasmas C1.1:6
Race track A3.0:1
Radiation sources B1.3:10
Radiation-enhanced processes A1.4:6
Radiative cooling E2.2:25
Radiative transitions D2.4:1, D2.4:3, D2.4:5
Radical–radical collision frequency B1.2:8
Radical–radical interactions B1.2:3
Radicals B1.2:1
Radio frequency (RF) sputtering A3.0:1, A5.0:1, A5.2:1, A5.2:4, A5.2:6
Raman shifting B1.3:11
Raman spectroscopy D0:6
Range E2.2:4
  calculations E2.2:6
  distributions E2.2:6
  projected E2.2:3, E2.2:9
  radial E2.2:6
  reduced E2.2:7
  spreading E2.2:6
  straggling E2.2:9
  tables E2.2:9
Rapid sweeping D3.0:2
Rate equation for pyrolysis B1.3:8
Rate of incorporation D1.2:1
Rate-of-rise B1.2:14
Re-deposition C1.0:8, C1.0:9
Re-sputtering A3.3:6
Reaction efficiency B1.4:19
Reaction products B1.3:8
Reaction rate B1.3:3
Reaction rate constant A2.3:2
Reaction sequence B1.5:1
Reaction temperature F7:7
Reaction velocity F7:9
Reactions between the precursors B1.3:8
Reactive gas flow control A5.1:6, A5.1:8
Reactive gas partial pressure control A5.1:2, A5.1:5, A5.1:7, A5.1:9
Reactive ion beam etching C1.2:1
Reactive ion etching C1.1:3
Reactive ion vapour deposition A1.3:2
Reactive ion-beam sputtering A3.3:8
Reactive magnetron sputtering A5.1:1, A5.1:5
  control of A5.1:1, A5.1:5
Reactive multilayer materials F7:2
Reactive pulse plasma A1.4:8
Reactive sputtering A3.0:2, A5.0:1
Reactor geometric C1.1:3
Real-time control D0:1, D0:3, D0:6, D1.2:1, D1.2:3, D2.3:3
Real-time data analysis D2.3:5
Real-time measurement D0:1, D0:2, D0:5, D0:7, D1.2:1, D2.1:2, D2.1:3, D2.3:2, D2.3:3
Real-time monitoring D0:2, D0:3, D0:6
Rear-view LEED D1.2:3
Recipes for recoating LEED screens D1.2:5
Reciprocal lattice D1.0:1
Reciprocal lattice D1.0:2, D1.0:3, D1.0:4
  rod D1.0:1, D1.0:2, D1.0:4

Reciprocating probe D3.0:2
Recoil implantation E2.2:23
Recoil mixing E2.2:23
Recommended cable sizes A1.1:5
Reconstruction D2.3:3
Redeposition C1.3:1
Reduced energy E2.2:5
Reduction reaction B1.0:6
Reflectance D2.0:5
Reflectance anisotropy (RA) D0:6, D2.3:3
  spectrometer D2.1:2
Reflectance difference spectroscopy (RDS) D2.1:1
Reflected neutral A3.0:9
Reflection ellipsometry D2.3:1
Reflection high energy electron diffraction (RHEED) D0:4, D0:6, D1.2:1, D2.3:3
  intensity oscillation D1.0:5, D1.2:2
  oscillation D0:5, D2.0:1
Reflection mass spectrometry D0:5, D0:7, D0:8
Reflection modification coatings B1.2:5
Reflectivity D2.0:3, D2.3:1, D2.3:3
Reflectometry D0:5, D0:6, D0:7, D2.0:1
Refractive index D2.0:2, D2.0:5, D2.0:6
Refractory metal source A1.1:1
Region of confinement A3.2:15
Relaxation lengths F5:11
Remote deposition B1.2:11
Remote or downstream plasmas E3.1:13
Remote reaction B1.2:11
Replenishment and replacement of evaporant charges A1.0:13
Residual carbon D2.4:5
Residual gas analyzer D1.2:1
Residual gases A3.4:4, B1.5:13
Residual resistivity ratio RRR_H A3.4:12
Residual stress A1.4:15
Resist C1.0:1
Resistive anode network D1.2:4
Resolution D4.1:4, D4.1:5
Resonance techniques D4.1:6
Resonant frequency D4.0:1, D4.0:5, D4.1:6
Respiratory damage A1.5:10
Reynolds number B1.4:11, B1.4:14
RF magnetron sputtering A5.0:2, A5.2:9
RF sheath heating A3.1:18
Rinsing E1.0:15
Rod flame spray A4.1:7
Rotated substrate holders A1.0:11
Rotating analyser ellipsometry (RAE) D2.3:5
Rotating disc reactor (RDR) B1.1:3, B1.1:8
Roughening F5:12
Roughness A3.4:11
  of thin film boundaries F5:10

Safety A1.2:4, A2.2:14
  considerations B1.3:10
Sample cell B1.3:12
Sample rotation D2.3:6
Saturating surface reaction B1.5:1
Saturation conditions B1.5:3
Saturation dose, ϕ_s E2.2:17
Scale-up B1.2:13
  modular B1.2:13
Scanning electron microscopy D0:2, D0:4
Scattered radiation F5:13
Scatterometry D2.5:3
Scratch resistant coatings B1.2:5
Screened Coulomb interactions E2.2:3
Secondary ion mass spectrometry (SIMS) D0:6, D2.0:5
Secondary-ion angular distributions A3.3:4
Sector-type A3.4:8
Segregation F1:2
Selected area deposition B1.3:1
Selection of evaporation material A1.0:12
Selective desorption B1.5:5
Selective-area growth A2.2:2
Selectivity C1.1:1, C1.2:2
Self-aligned pattern transfer C1.0:6
Self-bonding A4.1:13
Self-cleaning probe D3.0:2
Self-organization F1:6
Self-propagating high-temperature synthesis F7:1
Self-propagating reactions F7:8
Self-sputtering A3.2:2, A3.4:3
Self-sputtering yield A1.4:6
Self-sustaining reactions F7:1, F7:2
Semi-insulating polycrystalline silicon B1.4:32
Semiconducting silicides B1.4:36
Semiconductor D0:6, D2.1:1, D2.3:3, D2.3:6
  alloys D2.3:4
  epitaxial growth D2.3:4
  film growth D1.0:2
  films D1.0:1, D2.4:1
  heterostructures D2.3:3
  optical quality D2.4:1
  quality D2.4:7
  surfaces D1.0:2
Sendust alloys X5.0:11
Sensitivity to surface roughness D1.2:1
Sensor head D4.0:2
Servicing A2.2:14
Shallow impurity levels D2.4:1
Si film A3.4:15
Si trench etching C1.1:8
Si-based optoelectronics F6.0:1
Si_3N_4 B1.4:30
Si_xGe_{1-x} D2.3:4, D2.4:7
$Si_{1-x}Ge_x$ quantum well structure D2.4:6
Si, Ge and related materials C1.1:5
Si–Cl–H system B1.4:6
Si–Ge A2.0:22, F1:2

Si–Ge–Cl–H system B1.4:6
SiGe alloys B1.4:33
SiGe FET F6.0:12
SiGe waveguide F6.0:5
SiO_2 B1.4:29
SiON B1.4:30
Sidewall deposition C1.0:4
Sidewall passivation C1.1:6
Sigmund's theory A3.4:3
Silicon germanium F6.0:12
Silicon p-i-n detector D2.0:4
Silylated resists C1.1:11, C1.1:17
Single-shot reactors B1.5:9
Skin effect X5.0:9
Smoothing F5:12
Soft X-ray A3.4:21
  microscopes F5:13
  telescopes F5:13
Solid-source MBE A2.0:1
Source configuration A1.1:7
Source of radiation B1.3:9
Space charge neutralization A3.4:8
Spectrograph D0:6, D2.3:5
Spectrometer D0:5, D2.4:2, D2.4:7
Spectroscopic ellipsometry D0:3, D0:6
Spherical distortion D4.1:2
Spherical probes D3.0:4
Spitting A1.0:12
Spot shapes D1.2:2
Spot velocity A1.4:2
Spray/fuse A4.1:4
Spray+fuse A4.1:4
Spray torch A4.1:6
Spraying E1.0:11
Sputter etching C1.0:8
Sputter saturation E2.2:17
Sputtering A1.0:3, A3.4:2, B1.2:6, D0:3, E2.2:14
  of alloys A3.0:11
  control concepts A5.1:2, A5.1:5
  of non-conducting materials A5.1:8
  physical A3.4:3
  species A5.0:3, A5.0:5
  threshold energy E2.2:15
  yield A1.4:7, A3.0:10, A3.4:3, E2.2:14
Stability A1.2:7
Stacking faults A2.0:30, D1.0:3
Start-up transients B1.2:7
Static secondary ion mass spectroscopy (SSIMS) E3.1:2
Step arrays D1.0:4
Step coverage A1.0:11
Step density D1.0:5, D1.0:7, D1.0:8
Step edges D0:4, D1.0:5, D1.0:8
Step propagation D1.0:5
Step-flow growth A2.0:26
Stepdown transformers A1.1:6
Stibine, SbH_3 A2.1:13
Sticking coefficient A2.1:6, A2.2:2, A2.2:17
Sticking probability A3.4:4
Stoichiometry A3.4:21, D0:6
Stopping cross-section E2.2:4
  total E2.2:4
Straggling E2.2:3
Strain F2:2
  coherency F2:5
Strain relaxation D1.0:10
Strain-free windows D2.3:6
Strained layers F6.0:13
Strained-layer growth D1.0:10
Strained-layer superlattice A2.0:29
Strained SiGe alloy F6.0:1
Stranski–Krastanov mechanism A2.0:29
Stress B1.2:8, D0:2, D0:8, F2:1
  anisotropy D4.1:2
  coherency F2:1
  compressive B1.2:9
  interface F2:2
  intrinsic B1.2:9, F2:2
  tensile B1.2:9
  thermal F2:1
Stress-induced frequency shift D4.0:5
Strippable coating E1.0:15
Stripping E1.0:2
Stripping of photoresists C1.1:11
Stylus profilometer D0:5
Stylus profilometry D0:4
Substitutes for arsine and phosphine A2.2:16
Substractive pattern transfer C1.0:2, C1.0:5
Substrate deflection D4.1:1
Substrate displacement D4.1:4
Substrate distortion during deposition D4.1:1
Substrate heating A5.0:6
Substrate holders A1.0:9
Substrate rotation D2.0:1, D2.0:4
Substrate temperature A2.2:1, D0:7, D0:9, D1.2:3, D2.0:4, D4.1:7
Supercritical fluids (SCF) E1.0:6
Superlattices A2.0:30, F1:1
Supermodulus effect F2:3
Supersonic expansion mechanisms A1.5:3
Surface D2.3:3
  anisotropy D2.1:1, D2.1:2
  asperities D1.0:1
  chemistry of ploymers E3.1:1
  cleaning A3.4:15
  construction D0:1
  contact ionization A3.3:1
  dangling bond A2.3:2
  diffusion length D1.0:8
  disorder D1.0:4
  facets D1.0:4
  kinetics B1.4:17

migration D1.0:5
migration length A2.0:23
modification E3.1:1
morphology D0:1, D1.0:1, D1.0:5
perfection D1.0:1
photosimulated reactions B1.3:4
preparation E1.0:1
reconstruction A2.0:30, A2.3:2, B1.5:4, D0:2, D0:6, D1.2:3
roughness D0:6, D1.0:4, D2.1:1, D2.1:4, D2.5:4
segregation D1.2:3
smooth A3.4:25
step density D1.0:11
stoichiometry D0:4, D1.0:2, D1.2:3
structure D1.0:1
symmetry D1.0:1
temperatures D2.0:1
topography D0:4
vapour pressure A1.2:1
work function D3.0:1
Surface-sensitive D1.2:1
Symmetric reactors B1.2:9
Synchrotron radiation D0:4
Synthesis B1.0:10
System designed A2.2:3
System monitoring B1.3:15
System operation A2.2:12

Ta_2O_5 A3.4:19
TLV of AsH_3 A2.1:6
TLV of PH_3 A2.1:6
Target current control A5.1:6
Target electrode A3.1:12
Target poisoning A5.0:3, A5.1:1, A5.1:8, A5.1:9, A5.2:3, A5.2:5
Target voltage control A5.1:6
Temperature D0:2, D0:8, D2.3:3
at the film surface D1.2:2
effects D4.0:4
measurement D2.2:4, D2.2:7
Tensile strength F2:7
Tetraethyltin (TESn) A2.3:7
Texture D0:1, D0:3
Thermal barriers A4.1:15
Thermal cleaning A2.2:17
Thermal conductivity D0:2
Thermal cracker A2.2:10
Thermal diffusion F7:8
Thermal diffusivity F7:9
Thermal dissociation A2.2:1
Thermal emission D2.0:2
Thermal ionization A3.3:1
Thermal isolation A2.3:12
Thermal spike model A3.4:17
Thermocouple D0:7, D2.4:1
Thermodynamic calculations B1.4:6
Thermodynamics of deposition B1.3:7
Thickness D0:4, D1.2:1, D2.0:1, D2.3:2
control D2.3:5
non-uniformities D0:8
uniformity B1.4:14, D0:2
Thin film characteristics D0:1
Thin film deposition X3.6:2
techniques F7:1
Thin film head X5.0:11
Thin film hybrid circuits A5.0:1
Thin film interference effect D2.0:1
Thin film monitoring B1.3:15
Thin film transistor (TFT) B1.2:5
Thornton phase diagram A3.0:13
Threading dislocations A2.0:30
Three-dimensional growth D2.1:4
Threshold limit values (TLV) A1.5:10, A2.3:16
TiN A1.4:3
Time-dependent temperature distribution A1.5:4
Titanium dioxide (TiO_2) X3.5:2
Total integrated scatter D2.5:2
Total reflection F5:4
Townsend discharge A1.4:1, A3.0:2
Toxic gas monitor A2.2:15
Toxicity of the hydrides A2.1:5
Transfer width D1.2:4
Transients B1.2:8, B1.2:16
Transmission diffraction pattern D1.0:4
Transmission electron microscopy D0:2, F6.2:1
Transparent conducting oxides D0:8
Transverse optical phonons (TO) D2.4:6
Travelling wave reactor B1.5:9, B1.5:10
Triangulation D4.1:4
Tribological coatings A1.4:7
Triethylaluminium (TEAl) A2.3:5
Triethylgallium (TEGa) A2.1:7, A2.3:2
Triisopropylgallium (TIPG) A2.3:7
TRIM E2.2:9
Trimethylamine alane (TMAAl) A2.3:7
Trimethylgallium (TMGa) A2.1:7, A2.3:2
Trimethylindium (TMIn) A2.3:3
Triode B1.2:9
magnetron A3.2:23
reactors B1.2:11
sputtering A3.0:8
Triple probes D3.0:5
Trisdimrthylamino-arsenic (DMAAs) A2.3:7
Tuning the suppressor voltage D1.2:5
Twins A2.0:30, D1.0:3
Two- and three-layer masking C1.1:11
Two-dimensional (2D) disorder D1.0:2
Two-dimensional (2D) layer-by-layer growth D1.0:5
Two-dimensional nucleation D1.0:5, D1.0:6
Two-dimensional surface disorder D1.0:3
Two-dimensional unit cell D1.0:1

Two-layer resist C1.0:4
Two-photon absorption B1.3:4
Two-zone cells A2.0:5, A2.0:7
Types of reactors B1.0:2

UM-gun A3.2:11
Ultra high-vacuum CVD B1.4:24
Ultra-high vacuum (UHV) D0:2, D1.0:1, D2.3:3, D2.4:2
Ultrapure water E1.0:15
Ultrasonics E1.0:11
Unbalanced magnetrons A3.2:9
Uniformity C1.0:10
Unimolecular rate constant B1.4:15
Unwanted mixing B1.3:14
Up-to-air procedure A2.2:14
UV irradiation E3.1:12

V-series pyrometer D2.0:1
Vacuum arc A1.4:2
Vacuum environment A1.0:1
Vacuum pumping A2.2:5
Vacuum pumps A2.2:18
Vacuum ultraviolet photochemistry E3.0:9
Valence band D2.4:1
Valved cracker cells A2.0:7
Vapour condensation E1.0:11
Vapour phase epitaxy A2.0:2, B1.1:1
Vapour phase reactions B1.3:4
Variable angle of incidence C1.0:10
Velocity distribution of the ablated particles A1.5:3
Vertical reation cells B1.3:15
Vicinal surface D1.0:4
Video camera D1.2:4
Voltage index A3.2:1
Vortex rolls B1.3:14

W–Si–Cl–H system B1.4:8
Wafer temperature D2.4:1
Wavelength D2.4:7
  for excimer lasers B1.3:12
Wear resistance A4.1:14
Wear resistant coatings A1.4:1
Wet chemical etching E1.0:4
Wet etch C1.0:7
Wetting B1.2:5
White-light interferometry D2.2:3
Windowless system B1.3:10
Wire flame spray A4.1:3, A4.1:4
Woods' notation D1.2:2
Work function A3.4:13
Workpiece cooling E2.2:25

Xerographic photoreceptors B1.2:4
X-ray analysis D0:2
X-ray diffraction D0:3
X-ray optics A3.4:21
X-ray photoelectron spectroscopy (XPS) D0:6, E3.1:1

$YBa_2Cu_3O_y$ A3.4:21
Yield strength F2:7

Z-factor D4.0:4
Zero-flux contour A3.2:11
Zincblende-type semiconductors D2.1:1
ZnSe A2.0:2, A2.0:21, A2.0:22

For Product Safety Concerns and Information please contact our EU representative GPSR@taylorandfrancis.com Taylor & Francis Verlag GmbH, Kaufingerstraße 24, 80331 München, Germany

Batch number: 10731600

Printed by Printforce, the Netherlands